T0312613

Gina Mastantuono, Chief Financial Officer, ServiceNow
"Finally! *The Adventure of Sustainable Performance* is the ESG guide and an invaluable resource for CEOs, CFOs and any enterprise striving to meet stakeholder expectations. Stuart and Dean remind us that we must go beyond environmental responsibility and fulfill our corporate social and governance obligations. Embarking on an ESG journey is no longer a nice-to-have; it is a business imperative."

Jack Leslie, Former Chairman, Weber Shandwick
"I was suspicious when I read the authors' claim that 'this is not just another business and sustainability book.' The fact is that, for many CEOs, the constant drumbeat of stakeholders demanding sustainable business practices and 'shared value' has begun to give them headaches."
Dean Sanders and Stuart McLachlan have not only been on the frontlines of the sustainability movement, they've been in the boardrooms of companies working with business leaders who must pivot to new ways of doing business. This book will open your eyes to the enormous change that lies ahead – change that goes well beyond just commitments to reduce carbon emissions. Change that will truly transform industries and economies."

Cristina Gallach, Spanish sustainability expert,
former UN Under-Secretary-General,
and deputy Foreign Affairs Minister
"Uncertainty, disruption, complexity and accelerating change define this moment in time. Business leaders must deliver transformational change alongside positive impact and returns for investors. But in addition, they have to face up to their responsibilities to steer a different course for our endangered planet. Stuart and Dean's invaluable book takes ESG to the next level. They stretch ESG practices into new time horizons and to a destination they refer to as 'sustainable performance'. Precisely the adventure that all committed and responsible leaders have been waiting for! This book will act as a reliable and conscientious lighthouse in a uniquely challenging environment."

Dame Polly Courtice, Founder and Former CEO,
Cambridge Institute for Sustainability Leadership
"This book leaves the reader in no doubt about the fierce 'reality of now' and the turbulent times that lie ahead. More importantly, it is a wonderful call to arms for leaders who want to embrace the transformation that is required and seize the opportunities that lie within sustainable performance and value creation. It is filled with great examples of the dynamic, adaptive, courageous leadership that will be required, and offers invaluable guidance in understanding the language and developing the tools that will be needed in the next stage of their journey."

Santiago Gowland,
Chief Executive Officer, Rainforest Alliance
"In *The Adventure of Sustainable Performance*, Stuart McLachlan and Dean Sanders remind us of the power of leadership in steering our enterprises and economies towards a decarbonised and regenerative future. At Rainforest Alliance we share the conviction that partnerships between the private sector, NGOs and governments are key to accelerate the shift to what the authors call 'total value systems' at the scale required this decade. This is essential reading for any leader who wants to be part of shaping that future."

Paul A. Davies,
Global ESG Co-Head at Latham & Watkins
"A book that is both highly engaging and vitally important for business leaders globally. As Stuart and Dean rightly note, the business world has a tremendous opportunity to make choices that create sustainable, long-lasting value for society as a whole – *The Adventure of Sustainable Performance* is an excellent source for those wishing to do so."

'The wisdom contained in this brilliant book will
become seminal, and its impact, transformational'

—Henri-Claude de Bettignies
Emeritus Professor at INSEAD Business School

THE ADVENTURE OF
SUSTAINABLE
PERFORMANCE

BEYOND ESG COMPLIANCE
TO LEADERSHIP IN THE NEW ERA

STUART MCLACHLAN
DEAN SANDERS

This edition first published 2023

© 2023 Stuart McLachlan and Dean Sanders

All rights reserved. No part of this publication may be reproduced, stored in a retrieval system, or transmitted, in any form or by any means, electronic, mechanical, photocopying, recording or otherwise, except as permitted by law. Advice on how to obtain permission to reuse material from this title is available at http://www.wiley.com/go/permissions.

The right of Stuart McLachlan and Dean Sanders to be identified as the authors of this work has been asserted in accordance with law.

Registered Offices
John Wiley & Sons, Inc., 111 River Street, Hoboken, NJ 07030, USA
John Wiley & Sons Ltd, The Atrium, Southern Gate, Chichester, West Sussex, PO19 8SQ, UK

Editorial Office
The Atrium, Southern Gate, Chichester, West Sussex, PO19 8SQ, UK

For details of our global editorial offices, customer services, and more information about Wiley products visit us at www.wiley.com.

Wiley also publishes its books in a variety of electronic formats and by print-on-demand. Some content that appears in standard print versions of this book may not be available in other formats.

Trademarks: Wiley and the Wiley logo are trademarks or registered trademarks of John Wiley & Sons, Inc. and/or its affiliates in the United States and other countries and may not be used without written permission. All other trademarks are the property of their respective owners. John Wiley & Sons, Inc. is not associated with any product or vendor mentioned in this book.

Limit of Liability/Disclaimer of Warranty
While the publisher and authors have used their best efforts in preparing this work, they make no representations or warranties with respect to the accuracy or completeness of the contents of this work and specifically disclaim all warranties, including without limitation any implied warranties of merchantability or fitness for a particular purpose. No warranty may be created or extended by sales representatives, written sales materials or promotional statements for this work. The fact that an organization, website, or product is referred to in this work as a citation and/or potential source of further information does not mean that the publisher and authors endorse the information or services the organization, website, or product may provide or recommendations it may make. This work is sold with the understanding that the publisher is not engaged in rendering professional services. The advice and strategies contained herein may not be suitable for your situation. You should consult with a specialist where appropriate. Further, readers should be aware that websites listed in this work may have changed or disappeared between when this work was written and when it is read. Neither the publisher nor authors shall be liable for any loss of profit or any other commercial damages, including but not limited to special, incidental, consequential, or other damages.

Library of Congress Cataloging-in-Publication Data

Names: McLachlan, Stuart, author. | Sanders, Dean, author.
Title: the adventure of sustainable performance : beyond ESG compliance to leadership in the new era / by Stuart McLachlan and Dean Sanders.
Description: Hoboken, NJ : Wiley, 2023. | Includes index.
Identifiers: LCCN 2022056915 (print) | LCCN 2022056916 (ebook) | ISBN
   9781394177417 (cloth) | ISBN 9781394187263 (adobe pdf) | ISBN
   9781394187256 (epub)
Subjects: LCSH: Leadership. | Sustainable development. | Social
   responsibility of business.
Classification: LCC HD57.7 .M39596 2023 (print) | LCC HD57.7 (ebook) |
   DDC 658.4/092—dc23/eng/20221205
LC record available at https://lccn.loc.gov/2022056915
LC ebook record available at https://lccn.loc.gov/2022056916

Cover Design: Wiley
Cover Images: Elephant: © bartamarabara/Getty Images
Canoe: © suirey/Getty Images
Paper Texture: © Silmairel/Getty Images
SKY10048188_051823

Canoeing Under Elephants

How an Encounter with Elephants Inspired the Adventure of Sustainable Performance

In writing this book we wanted to express a topic often portrayed as dry, and at times even apocalyptic, into a subject that is more evocative and adventurous. The 'Canoeing under Elephants' metaphor frames the entire book but we have tried to illustrate the key messages with the use of images. Some of these are metaphorical, others pull on history, while others are real life stories from fellow guides and adventurers. All have helped us to stay true to the reality of the existential crisis that we face, and at the same time, inject excitement, pragmatism and hope into the opportunity that this moment in history presents.

The authors' proceeds for this book will be invested into African wildlife conservation.

Canoeingunderelephants.com

To Albert Freese. Without his entrepreneurship, conviction and belief, none of this would have been possible. Danke.
Dean Sanders

To Chris Cole and Malcolm Paul. For taking risk, for believing in me and for allowing me to learn from my mistakes.
Stuart McLachlan

To our families for sharing and supporting in this adventure.
To our fellow Anthesians (fellow Canoeists) for creating value, your fun and your impact.
To our clients for your spirit of enterprise.
Dean Sanders and Stuart McLachlan

Our thanks. . .

To Mike Thompson. For pulling it all together with unfailing enthusiasm, and for your incredible eye for detail.

To Sarah Coope and Jamie Burdett. For wrestling with us through the early concept phase, for keeping us real, for bringing us back whenever we wandered off, and for your brilliant research, creative ideas, and imaginative thinking.

To our fellow Anthesians. For your pioneering instincts, your innovation, your solutions, and your scientific integrity. For reminding us why we matter as individuals, as teams and as a business.

To our interviewees: For your honesty, your bravery and your clear sightedness. For being willing to impart your wisdom and experience to help those yet to embark upon this exciting adventure.

To our families. For putting up with our absence and, when we have been present, for being gracious in your feedback as we have exposed you to those early drafts.

Contents

Foreword

Henri-Claude de Bettignies
Aviva Chair Emeritus Professor of Leadership and
Responsibility and Emeritus Professor of Asian
Business and Comparative Management at INSEAD

It is unusual to come across a business book that starts with adventure and the intriguing metaphor of 'Canoeing Under Elephants'. Then the meat – a solid diagnosis with realistic aims and, ultimately, a well-defined plan of action that turns into an urgent call to arms!

In a nutshell, this is a great book that supplies a vitamin boost to the hope we all need to escape the pessimism of a dystopian world or from the one returning us to the Amish candle or the nightmare of a quasi-inhabitable, dry planet that will be home to many of its 10 billion inhabitants by 2050 or any one of the many other gloomy scenarios for our future.

The elephants watching us by the side of the river are actually already in the room, the product of our irresponsible behaviour, our abuse of planet earth through hubris and the savage exploitation of its finite resources. They are the result of our broken partnership with nature.

Using their highly original metaphor throughout the book, two 'unusual consultants', Stuart McLachlan and Dean Sanders, take us on an alternative, potentially optimistic scenario, if we are willing not to look elsewhere when our house is really on fire. 'The Adventure of

Sustainable Performance' describes the true state of our fragility and throughout beats the drum sounding today's very real dangers. But it also lays out the competences and actions necessary for us to navigate quickly a way out of the dark threat of climate change. The authors are not whistle blowers – it is too late for that. Rather they sound the tocsin to signal that the fire is here. Now we really must hear it.

We do have solutions and the authors show us the road, albeit a challenging one, that we must take. We must urgently engage in a long overdue paradigm shift. The dominant model that has guided our achievements so far has also led us into the present catastrophic situation. It is now obsolete. The media illustrate this daily, notably through their very gloomy reports of the seasonal COP meetings and the devastating human and ecological consequences of increasingly frequent extreme events, while scientists throughout the world provide plentiful reliable data to back up the evidence.

However, in the corporate world, a few still feign ignorance and continue to ignore this noisy message. Some find comfort in ticking the boxes of compliance while others indulge in greenwashing. But today, a fast-growing number of business leaders are well aware of the scale of the crisis and have developed their long-term vision to grab new opportunities with imagination and agility. They realise that 'business as usual' is no longer an alternative, it is suicidal. Hence they are willing to challenge the 'strongholds' of their present position and take the road less travelled. They are shifting from the shareholder model to the stakeholder paradigm, a transition they *finally* see as a categorical imperative.

The great value of the authors' argument is that – based on years of experience with the complex reality of corporate transformation – they actually tell us how to steer our canoe in this uncertain, high risk, fast-changing environment.

They tell us how to escape the fallacy of endless growth, profitability at all costs, destructive competition and corporate myopia. They also warn us that an individualistic mindset cannot solve systemic challenges, and that passing the buck to the government, from government to business and business to the individual consumer is neither responsible nor effective.

In order to show the limits or ineffectiveness of compliance-only approaches, of corporate social responsibility and short-term ad hoc measures that do not consider the 'total value system' of the firm, they make crystal clear that the ecosystem of the whole supply chain must be based on a realistic, long-term view that includes all stakeholders. Many examples are given, from large multinational corporations led by enlightened CEOs, to family firms that take the long view in managing transition and succession, and start-ups that are willing to risk bold innovations. They illustrate how realistic business leaders (for example B-Corporations) who are purpose-driven and take alternative roads toward transformative sustainability, have seized upon the opportunities brought by the energy transition imperative.

This is a book of distilled experience that not only makes explicit the many obstacles, defence mechanisms and resistance to change (the 'strongholds'), but also exposes the many positive factors that can be leveraged to facilitate the long overdue paradigm shift. One of these is illustrated by the growing interest in, and pressure for, ESG among investors. This starts to have an impact on corporate governance, it induces boards to broaden their vision, to better assess complex interdependencies among stakeholders, to adopt a more holistic view of the organisation and its supply chains, and to monitor more responsibly the compensation packages for top and C-suite executives. Such investor interest will also act as a stimulus for research in order to accelerate the needed improvement of ESG measurement methods and tools.

My enthusiasm for this book stems from the fact that it clearly illustrates – through so many examples and well-chosen quotes – the many messages I have been conveying over many years of teaching senior executives. Thirty years ago when discussing the issue of 'responsibility' with business leaders I used to insist on the necessity to clearly identify and manage all externalities, including the impact of our actions on the 'birds', which I used as a proxy for nature. Polite smiles were the usual reaction. What I wanted was to broaden executive awareness of the many interdependent parts of the corporation as a complex ecosystem, and the leader's responsibility for their management. I did not fully elaborate on the supply chain that ESG now integrates so well and I am delighted to see how Stuart McLachlan and Dean Sanders brilliantly articulate that imperative for sustainable performance.

In the same way, while trying to broaden awareness, I was also keen to sharpen it, by pushing leaders to become more clearly aware of the filters (biases, prejudices, stereotypes) that blur their perception of the world and of the organisation and which too often locked them into the defensive position that the authors identify as a 'stronghold' blocking a possible paradigm shift.

To move to a paradigm that integrates all stakeholders – one that ESG requirements will eventually render compulsory – will be a painful process for many, particularly in sectors such as the fossil fuel or automotive industries where lobbying and protective defence mechanisms are so powerful.

It will require industries and services to integrate the whole value chain as an ecosystem, to rethink priorities and the time dimension while avoiding any temptation of greenwashing.

Government regulations, media pressures, investor requirements, younger generations' expectations, civil society noise and tomorrow's rebellions by the socially excluded will push for change. It will be adding to the pressure from extreme events and the ubiquitous impact of climate change on nature, on finite resources availability, on biodiversity damage, and so on. ESG will then become mainstream and illustrate that, finally, the paradigm shift that is already a categorical imperative is actually happening. Courage is and will be in high demand.

The leader who does listen to the authors' experience and reasoning should not be obsessed or paralysed by uncertainty and fear of the unknown, but driven by confidence in their own values, clearly articulated using purpose as a compass to seize opportunities along a possibly long and treacherous river. Such leaders are, and will be, in great demand!

Will business schools contribute effectively to develop this leadership profile? To create leaders who care for all stakeholders and are willing to act as responsible change agents to transform our current corporate world's mindset for the benefit of their grandchildren? The jury is out!

If all managers could be required to read McLachlan and Sanders' book, the wisdom hidden in its analysis would become seminal and the authors' wise proposals would certainly inspire and drive action. Hence it is a modest, but extremely useful, contribution to try to avoid the worst and build a better world. Before it is too late!

Prologue: Encounters with Elephants

Several years ago, Anthesis co-founder and co-author of this book Stuart McLachlan had an experience with elephants that had a profound impact on him. When he shared his story with the rest of us at Anthesis, we realised that this was a story that captured the essence of the challenges that we, as humanity, currently face.

Day 1

A number of years ago, my wife Louisa and I travelled to a remote region of Africa on the Zambezi river. We were excited for the experiences and adventures that lay ahead as we set out on a walk through the bush with our guide...

As we walked, the guide pointed out plants and insects, as well as birds flitting between the trees. The environment was peaceful and unthreatening. Suddenly, with an almighty crash, a large bull elephant hurtled out of the bush in front of us. He looked startled as he registered us in front of him, and we were equally shocked. The pause lasted barely seconds before the elephant ran off.

'That was close!' I exclaimed to the guide as the elephant's thundering run took him further from us. The guide didn't look in my direction, his eyes were still keenly trained on the elephant. 'Stand behind me, stay still, say nothing,' the guide said in a serious and commanding tone,

but with an undercurrent of fear. We complied immediately, now also watching the elephant closely. As we stood behind our guide, the elephant turned and charged.

There was a sudden explosive shower of noise, movement and aggression as the bull elephant trumpeted, waved his head and ears and ran towards us at speed. The guide kept repeating his instruction, 'Stand still, say nothing'. Feeling helpless to the power of the elephant and standing rock still, we watched as our guide raised his gun above his head and spoke firmly to the elephant.

'Voetsak!' he commanded (an impolite Afrikaans term for 'go away', in case you're wondering). Dust and detritus swirled around us, we were frozen in fear and cognitively overloaded. The seconds seemed to stretch for minutes, but, as abruptly as the charge started, it stopped. The elephant had assessed the situation, calmed down and returned to the forest, leaving us with pounding hearts and a very different perspective on our environment.

Day 2

The following day, we set out with the same guide for a very different kind of elephant experience. This time the intention was to approach not one, but a group of those mighty beasts from the river, in a canoe. The canoe was small, open and felt very vulnerable. After the heart-racing intensity of the previous day, we naturally expressed some strong reservations (and yes, some fear) about reapproaching what was by then a proven source of life-threatening danger in our eyes.

Paddling gently towards the very creatures that had been so terrifying just a day earlier seemed both counterintuitive and irresponsible. Words with the guide were exchanged, our nerves were jagged, and we felt vulnerable at the thought of, once again, putting our lives in the hands of untamed nature. Our guide, however, was calm and relaxed. He explained that as the elephants have never experienced danger from the river, they wouldn't see the canoe as a threat. Therefore, the canoe and its passengers, us, would be totally safe.

Reassured, we boarded the canoe with our guide and tentatively paddled towards the elephants who were standing along the bank of the river, just a few metres above the gently flowing water.

The canoe trip became the very best kind of adventure. All our previous fear was replaced with wonder and joy as we marvelled at the elephants standing just above us. It was the kind of experience that defines lifetimes. Amazingly, in the canoe, we were completely part of the elephants' environment. As our guide promised, we presented no threat and were able to pass safely underneath the elephants as they ate plants from the river. Water was dripping onto us from the mouths of the elephants and we could hear their snorts and rumbles as they ate and interacted within the herd.

Drifting slowly beneath the elephants, absorbed by our interactions, we felt a deep sense of peace and connection to these incredible, magnificent animals and the nature they inhabit.

We are at a pivotal moment in time. Both today and the future are filled with change, ever-increasing complexity, and uncertainty beyond what we, the authors, have ever known. When Stuart shared this experience that he and Louisa had in Africa, we were captivated. We felt that this story captured the fear many business leaders face in today's uncertain world, as well as the trepidation with which they approach new opportunities. We also realised that, were it not for the guide, Stuart and Louisa would never have climbed into the canoe or had this once-in-a-lifetime encounter with the elephants.

This is not just another business and sustainability book. This is a book that seeks to get under the skin of the reality of what leaders and businesses face today: a myriad of demands and expectations. In addition to the usual challenges of running a sound and profitable business, leaders are expected to take decisive action in response to both the threat of climate change, and the complex societal challenges that sit alongside it.

This can feel like an impossible balance. Rather than inspire determined action, it can cause us to freeze, uncertain of what to do, beyond following a maze of compliance. Compliance has its place, but in this moment, in which everything is shaking, compliance is not enough. This book is for those who know that things need to change but are puzzling how or where to begin. We want to inspire you through our own experiences, as well as through the stories of businesses and leaders who are reapproaching the elephant 'charge' in a different way. We aim to show you how this can open up the possibility of reimagining our future and enable you to make choices that will be more prosperous for all.

As we go through the book, we will make reference to Day 1 and Day 2 to contrast the two responses, and to frame the adventure.

Day 1 – The Charge

This is the reality of now; the challenges we face both as business leaders and humanity. The Charge represents the complex web of concerns, pressures and uncertainties that can cause us to freeze, to not take decisive action, and to blindly comply, rather than innovate. It is the uncertainty of an unknowable outcome that can lead us to hunker down and protect what we know rather than step into opportunity.

Day 2 – The Reapproach

The Canoe represents something different – opportunity, possibility and hope. It represents what can happen when we approach a fear or challenge from a new perspective. Despite the fear of the charging elephant and the vulnerability of the canoe, Stuart and his wife trusted their guide and embarked on an adventure that turned out to be a calm, safe, experience and a life-changing moment. Our hope is that this book will take you on a similar adventure.

The Adventure

The environmental and societal challenges we face today are like nothing we have ever seen. Many of us now know that business as usual is no longer an option, but the alternative paths of action seem unclear, likely expensive and, quite frankly, difficult. We want to reframe this idea, to show that while there is great uncertainty and even elements of danger in this moment and the future ahead, for the leaders willing to try, this is also a great adventure. We have the opportunity now to be part of something that is as exciting as it is hopeful, where we can create business ecosystems within nature's ecosystems that will enable us to experience a safe, harmonious and prospering future.

What qualifies us to do this? Well, in different ways as individuals and leaders we see ourselves as essentially Day 2 people. Both of us (Stuart and Dean), are entrepreneurs at heart who have left behind 'successful' careers in large multinational corporations to create new businesses focused on impact and sustainability. We have both confronted the binary choice of business, life or career 'as usual' and the alternative creation of something new and purposeful. Stuart co-created the Anthesis Group in 2012; Dean founded GoodBrand in 1997 and the two companies came together in 2019. We have experienced the sensation of branching out and embracing risk and return. What's more, we both feel privileged to have been able to follow our hearts as well as our minds and to pursue our passion for and commitment to the Sustainable Performance agenda. Each of us has consulted and advised the kinds of senior leaders who will lead the way to Day 2 and reapproach sustainability as an adventurous journey of impact and performance.

Introduction

Painting a dystopian picture of the future is the domain of many writers on climate transition and sustainability. We want to take you in an alternative direction, but to do so, we need to ground this book in the truth. To bring this to life, we decided to envision the world in 2050 according to emerging predictions, models, and assumptions. However, in doing so, what we discovered is that much of this 'future thinking' has already come to pass. We have therefore presented a picture that reflects what is happening during this 'decisive decade', so called because what happens in these 10 years is expected to decide what happens over the next thousand years. We are at a pivotal point in history, we need to face up to the actuality of it, and so we have called this section 'The Reality of Now'.

The Reality of Now

It is business as usual (BAU). We've ignored the warnings. We are entrenched in the model that has generated unprecedented wealth for much of the world's population for the last 80 years. The ways of the prevailing era remain dominant. But now we are facing a threat that appears unstoppable. Despite desperate pleas, leaders are yet to make the pivot needed to avert this existential crisis.

The global population has reached eight billion, and is heading for ten billion by 2050. There are many regions that are experiencing extreme weather impacts on a regular and sustained basis. Flooding and aridity are impacting agricultural supply chains and the sourcing of raw materials for manufactured products. Urbanisation is increasing. People are moving to cities to get work and to raise their standards of living as global climate change impacts bite. As a result of this migration, approximately 700 million people now live in coastal cities that face increasing threat from flooding, and US$1 trillion investment is estimated to be needed to protect cities from sea level rise. Real estate in parts of the world is becoming uninsurable.

Much of the population growth is in emerging economies – South East Asia, India, South America, and China. Tens of millions of climate migrants in coastal regions and the tropics are moving away from non-viable areas. Their crops have failed, or their homes are underwater. Conversely, other parts of the world have become uninhabitable because of water scarcity. This mass migration is accelerating.

The rise in urban living has fuelled a growth in consumerism sustained by increasing income and lifestyle changes. Demand for products and materials has increased due to population growth. Businesses attempt to succeed in the 'take, make and throw away' model that drove the consumerism bubble of the preceding 80 years. However, finding materials is challenging as resources become increasingly scarce in supply chains vulnerable to disruption from externalities. Decades of sourcing the raw materials to supply the demand for these products has resulted in significant reductions in natural habitat and biodiversity. The impact on some species is now irreversible. Many species have already become extinct. We are in the planet's sixth mass extinction, but the first driven by human activity.

The world knows that peak carbon emissions must occur by 2025 and that emissions need to be cut by 50% by 2030, by 50% again by 2040 and that we need to achieve net zero by 2050. If these targets are missed, we may face a catastrophic 3–5 °C of warming. Some leading scientists are suggesting that this now looks inevitable.

The per-head 'material footprint' of developing countries has increased by more than double. We are still heavily dependent on plastic to meet the consumption and production needs of the world's population. There are issues with material leakage – every minute there is an equivalent of four garbage trucks of plastic material discharged into the oceans. By 2050, at the current rate of disposal, it is predicted that there will be more plastic than fish in the world's oceans by weight. Two planets would be required to support the ongoing current demand within existing systems and economic models, with much of the developed world needing an equivalent of over five planets.

It is clear that the notion that nature can be tamed, that ecosystem services will be provided for no cost, or that nature and planetary based externalities will remain stable, is flawed.

Already at the periphery of this agenda, we see enlightened leaders thinking outside the box or creating new boxes within which to operate. They consider sustainability to be an essential business attribute and a source of value creation. They regularly change their business model, investing in supply chain partnerships and understanding the purpose-driven values of their customers. Adaptability is being seen as the new 'superpower' within the leadership community. Increasing evidence exists that those businesses operating in a model that creates shared value for the environment and society deliver superior financial returns for investors.

These examples are rare, but there are glimpses of new, durable business models emerging that place purpose at the heart of their mission. These models focus on stakeholder collaboration and on delivering sustainable and resilient performance through value cycles. They oppose traditional supply chains that seek leverage through the imbalance of buyer power over supplier power. Capitalism remains the driving force in global economies embracing the power of free trade. However, a less myopic and more responsible form of capitalism is increasingly talked about by brands and customers who can see the brokenness in the current system.

Despite the evidence underpinning this model of value creation, it isn't yet achieving traction in mainstream global markets. The competing forces of human innovation, with the intransigence of global institutions and short termism in political and business leadership tenures, is resulting in paralysis or a box ticking form of compliance. The world recognises it cannot sustain its population by continuing to operate the way it always has, and yet for many there is a reluctance to break the status quo. It's now too late to avert catastrophic global impact for much of the world's population.

This book is for business leaders who understand these dynamics. As authors we also understand where those leaders stand today and the demands made of them. We understand the contractual, fiduciary, and fiscal responsibilities and accountabilities that come with the job. So, this is not a book about sustainability performance, it's a book about *sustainable performance*. It's a guide for ensuring that the asset bases of businesses of different types and sizes can avoid becoming stranded, but can rather be redirected, redeployed, and reinvented to create new forms of value in a more sustainable world. It is a book about the opportunities in strategic and creative asset transition.

Reality in a 'Post-Truth' World

'The Reality of Now' is a tough read. It's stark and a jolt to action for some, while for others it is paralysing. We want to go to where you are. Whether you are up for action and excited by the opportunities presented in this disrupted world or weighed down by the burden of demands requiring you to be ever more compliant. Wherever you are positioned on this subject, we aim to provide you with clarity on what is the art of the possible in Day 1 and Day 2.

What this is not, is a book that sets out to persuade you of the reality with the scientific evidence or frighten you into a state of paralysis or panicked response. We are rooting this adventure in truth – we've got to get real – but our primary aim is to reveal glimpses of an alternative future that is hopeful and positive, a future that we are all empowered to create.

End of an Era

With as many as 60 million dead at the end of the Second World War, the destruction was incomprehensible. At this point in time, it was hard to contemplate going back to business as usual. Perhaps this is why 1945 is often referred to as 'Year Zero'. The new era was being rapidly designed as leaders became the architects for the era that was to come. The future was being chosen in this transitional space where the institutional and systemic strongholds dominant in the previous era were destroyed. Dynamism and the adaptability to reimagine the way the world worked supercharged the creative spaces that were freed up in this place of co-liberation.

Richard Rohr, author and theologian, refers to this transition zone as 'liminal space': 'Where we are betwixt and between the familiar and the completely unknown ... That's a good space where genuine newness can begin. This is the sacred space where the old world is able to fall apart, and a bigger world is revealed. If we don't encounter liminal space in our lives, we start idealizing normalcy' (Rohr, 2004).

It is this liminal space that we are entering as we come through a pandemic that locked down the world, and the world's economy, and as we contemplate the fact that we don't have access to two planets, only one. As the impact of climate change becomes alarmingly evident, and the existing business and economic models that created so much wealth in the previous era start to shake, we find ourselves at a new 'Year Zero'.

Throughout this book we will refer to the 'old era' as the one we are moving from. The transition zone and 'liminal space' describes where we are now, and the 'new era' is

where we are transitioning to. We recognise, though, that
for many of you the 'old era' is the current 'reality of now'
and that you are setting out into Day 1. For you the change
is ahead but it's imminent.

This book is framed in three parts: Part I – 'Day 1':
The Charge; Part II – The Campfire; and Part III – 'Day 2':
The Reapproach. These are stages in the journey to a future
that is reimagined and filled with opportunity. We are not
talking about an unachievable vision of utopia, but a reality already taking shape and an adventure that many have
already embarked upon.

Part I – 'Day 1': The Charge

Here we navigate the looming challenges leaders and businesses face in the current context. *The Charge*, of course,
represents the overwhelming power and uncertainty of a
natural world pushing back, but also the pressures, seen
and unseen, that cause us to freeze, resist, or retreat into
our current operating systems and mindsets.

The Rise and Fall of Strongholds

We are at the end of an era and not yet in the next. We are
in a period of rapid design change driven to a large extent
by the climate transition and by nature seeking to restore
its rightful position as our partner in any future we wish to
build. But with building something new comes destruction
of parts of the old.

Mark Sayers refers to the idea of strongholds in his
book *A Non-Anxious Presence* (2022). These are structures
that have been introduced to bring about security, stability,

and prosperity. These societal structures 'shape the thoughts, feelings, desires and attitudes of those living in that time. They form the places we look to for solace, strength, hope, and comfort. They create buffers for our anxiety. Their legitimacy is derived from making us feel safe and implicitly promising to give us an unfettered environment to pursue our best lives.'

Historians will point to the development of tribes to cities, then to regions, all of which aspire to become strongholds with these virtues. Storing more harvest (now economic wealth), strengthening the walls and the defence systems, and providing more prosperity for the inhabitants within the stronghold, is typically a preoccupation for leaders.

Strongholds are not inherently bad, we build them instinctively. But many have become so strong and immovable, reinforced, and rooted down, that change and transformation within them seems impossible. Many organisations and companies have also replicated a stronghold mentality. Much of the security within the company stronghold relates to an assumption that the environment will be tamed and that there will be ongoing stability in the climate outside the stronghold. The aim is to neutralise the impact of externalities and create calm and safety in our strongholds. Conversely, the places outside the strongholds are often seen as danger zones.

In a liminal space, strongholds become threatened as everything is changing. The natural response of a leader or group is to strengthen the walls and defence systems, to double down on what they have, to cling onto and protect the status quo at all costs. If you are enjoying the stability and security of these existing safe zones, then being told you need to step outside them into the danger zone, and embrace the uncertainty of the new era, is uncomfortable. Many don't want to accept the truth that their strongholds will be destroyed or changed beyond recognition as this era ends and the next begins. Instead, they choose to follow the rules, which are, generally, to comply, incur the cost of that compliance, and hope to get back to business as usual.

Conversely, we shall explore in greater detail later on how the shaking and potential destruction of certain strongholds can actually allow us to reapproach the challenges we face and provide opportunities to design new places of safety that represent shared value, social cohesion, inclusivity, resilience, and a place of 'total value creation'.

Systemic strongholds are many and complex, expanding beyond the world of business, and we will explore how interconnected these are and how they are all shifting in this liminal space. Though this book is primarily aimed at business leaders, we cannot forget that businesses sit within the dynamics of a complex, messy world governed by leaders from many different institutions and cultures.

We write this book at the end of the pandemic. War has returned to Europe. Energy prices are soaring and many families even in the wealthiest countries of the world will be choosing between energy or food this winter. The shaking of strongholds is growing in intensity, as is the desire to strengthen the walls and seek refuge.

Our challenge is whether the protection of the status quo, through instinct, hidden fear, or stubborn determination, can continue to keep us safe from the perceived danger beyond. Is doubling down now really in the best interests of our employees, our investors, other stakeholders, and us as leaders? Are you missing out on the opportunity to participate in the invention and development of the new era by doing so?

The Charge refers to the shifting tides of change with which businesses must grapple. An event which requires us to reconsider our responsibilities towards all stakeholders, where leaders operate across their entire business ecosystem.

Of course, there will be some businesses, ideologies, and institutions that pass more peacefully from one era to the next, but many strongholds will not withstand the Charge of climate change and its resultant impacts. Our concern is for the leaders who choose to ignore this, miss this, or – worse still – attempt to stretch past truths into a place of future fantasy, a tendency driven by the desire to survive a short-term tenure or to replace anxiety with comfort. The longer the fantasy realm continues to be the perceived reality, the more explosive the potential destruction is likely to be.

In the long term no-one will benefit from fantasy masquerading as truth. We want to call out the threat of the Charge in Day 1 and the opportunity of the Reapproach in Day 2, and just as we saw in the prologue, identify guides with whom you can experience the adventure. This will be a period when the world will continue to play at denial or talk more of ruins. However, the prosperous leaders, the leaders who will command followership, will see foundations, foundations upon which the new era will be built.

The Next Industrial Revolution

Many commentators are saying that we are on the cusp of the next industrial revolution driven by the climate transition. As Amory Lovins of the Rocky Mountain Institute says: 'In this decade, a rich stew of new technologies, materials, design methods, financial techniques, and business models, along with smart policies and aggressive investments, could revitalize, relocate, or displace some of the world's most powerful industries.'

This is not a time to hunker down in a sandcastle fort standing in the way of an incoming tide. It's time to step into the design work of the new era and get building.

The Fortune 500 companies represent broadly two-thirds of the US GDP with US$16.1 trillion in revenues, US$1.8 trillion in profits and US$22.6 trillion in market value. They employ approximately 30 million people worldwide and represent some two thirds of the US economy (Fortune, 2019). As of the autumn of 2021, 38% of the Fortune Global 500 had made commitments to reduce emissions by 2030, and 25% had adopted net-zero targets – up sharply from previous years.

Committing to and, most importantly, achieving net zero (as we will unpack) will demand a refocus on stakeholders, rather than maintaining an approach that focuses primarily on shareholders. This move from a shareholder to stakeholder capitalism model represents the shaking of an established stronghold and the construction of a new shared value model and inclusive form of doing business that has the potential to deliver superior returns.

In his open letter to investors in January 2022, Larry Fink, chairman and chief executive officer (CEO) of BlackRock, emphasised the importance 'in today's globally interconnected world' of companies creating value for and being valued by all stakeholders. This is not to detract from the importance of profit and profitability, which will remain a marker of success, but actually 'it is through effective stakeholder capitalism that capital is efficiently allocated, companies achieve durable profitability, and value is created and sustained' (Fink, 2022), thus creating long-term value for shareholders.

The Importance of Stakeholders in Value Creation

So, profit remains the oxygen that businesses need to breathe, the fuel that allows them to pursue purpose. But we shall invite you to reflect on why and how profit will be made in the new era. Foundational to the design will be a move to creating value across the entire value chain, what we call Total Value System integration and, in so doing, embracing all stakeholders.

We expect the relationship with stakeholders to change and we shall unpack what this looks like for the business and economic models of the future. For now, we want to introduce the importance of looking across a wider stakeholder ecosystem by looking at three key stakeholders.

- *Suppliers* – Achieving net-zero requires companies to take their supply chain with them. This means that companies are increasingly being held accountable for the net-zero ambitions of their suppliers. They can only meet their net-zero targets if their suppliers meet theirs. However, suppliers are often outside the stronghold. They sit on the other side of the procurement department where the current relationship is determined by the amount of buyer leverage that can be applied over supplier leverage with the aim of reducing input costs.
- *Employees* – If we are to deliver transformational change by 2030, employees will need to behave differently. Behaviour change programmes will need to be more ambitious than the average incremental nudge expectations of the old era.
- *Customers* – Now let's consider the relationship with customers, often seen as downstream and, again, typically outside the stronghold in the old era. Existing business

models are based on linear forward logistics. They rely on little or no responsibility after the point of sale. In the next era, we will see producer responsibility come to the fore, where companies will be responsible for products at end of life or, more likely, there won't be 'end of life' as products will be made in such a way that they can be recovered, dismantled, recycled, or reassembled. This circular economic model has spawned huge activity in secondary markets. Companies that are finding a way to reinvent value chains with reverse logistics are thriving by providing the market with quality, cost-effective, and sustainable products that present lower risk across the value chain and stronger long-term returns for investors.

We shall explore many other stakeholders as we move through this book, but by highlighting these three we get a sense of the changing relationships and the lift necessary across the entire value chain to achieve the net-zero commitments cited above – targets that are not optional if we want to direct ourselves away from the future anticipated earlier in this introduction.

Part II – The Campfire

The Campfire on the evening between the two days is about taking stock, reflecting on the reality of the Charge, the threat level, and the transition zone we need to step into. We will frame a new destination, the opportunity and hope related to it, and also the assets and team that you need to navigate the transition.

Many managers and leaders have been trained to assess a new situation, work out what needs to be done and what boxes need to be ticked to ensure compliance in accordance with the previous era. Many business school educational programmes, such as MBAs, were set up just a few years after the Second World War. They have still not stepped up to embrace the impact of sustainability on the way business is conducted. 'Schools are trapped in a business-as-usual mentality, cautiously seeking a balance between the need to update and the perceived need for the status quo. While businesses are advancing, business schools have some catching up to do', says Giselle Weybrecht in her *Financial Times* article in January 2022.

These leadership and management training programmes have conditioned many people to work to meet the expectations of shareholders and boards of directors, which stifles their ability to do things differently and disrupt. Change is often seen as something that is incremental, gradual, and linear. It is something businesses can apply their powerful models of efficiency to. Their corporate mindsets seek to find the most cost-effective way of achieving compliance with business-as-usual rules and formulae. Rules as fashioned by the textbooks of the old era.

We aim to challenge this thinking. The change that will be experienced in this liminal zone will call upon adaptability and innovation. Just as Henry Ford, founder of the Ford Motor Company, resisted the customer's call for a faster horse, so will we need to encourage the experts in the old era not to get in the way of the novices in the new. The liminal zone will call out the pioneering instinct in our leaders and reframe the agenda from cost of compliance to an investment proposition against which we should expect a return.

The World Economic Forum (WEF) *Global Risks Report 2022*, which calls on the views of 12000 leaders from around the world, found that climate risks dominate global concerns. As the world enters the third year of the pandemic at the time of writing, the top, short-, and long-term risks are perceived to be extreme weather and climate action failure. As far back as 2015, political leaders from over 190 countries recognised the consequences of staying on this trajectory when they signed the Paris Agreement at COP21. Less than a decade on, we see the effects of a world straining under the pressure of climate stress, with the word 'unprecedented' being used every few weeks to describe extreme climatic events.

Perhaps this is the reason why investment in climate tech is continuing to show strong growth as an emerging asset class. The PwC *State of Climate Tech Report 2021* reports a total of US$87.5 billion invested over the second half of 2020 and the first half of 2021. This is a 210% increase on the 12 months prior. Climate tech now accounts for 14 cents of every venture capital dollar (PwC, 2021).

The acceleration of enterprise at the nexus of climate and tech is expected to grow exponentially as companies, cities, and governments seek transformational interventions to deliver the essential pivot necessary to perform rather than perish in this decisive decade.

As venture capitalists and tech entrepreneurs move aggressively into this space, there is a race to capture capital destined for impact investment or assets with high Environmental, Social, and Governance (ESG) scores. Those businesses birthed into Day 2 and seeking capital will be experiencing different challenges to those long-established entities facing the Charge – the canoe adventure for a

100-year-old steel manufacturing company will look very different to that of a climate tech start-up.

Against this backdrop of one size not fitting all, we'll explore what this means for leaders and introduce the guides in this context. We'll demonstrate how calm, clarity, and experience can help leaders navigate the way forward as well as delving into the typical attributes that need to be developed across leadership teams.

For the avoidance of doubt, we are under no illusion that stepping into this liminal space, a danger zone for many, will generate anxiety. The decisions we take now will shape the future for many generations to come. Leaders will need guides who have been to the destination and can support them in finding the best route. These guides (and hopefully this book) will set out to do two things. Firstly, to persuade you that this transition is inevitable; that there is no choice but to step out of the walled fortresses, built in the last era, into a transition zone. Secondly, and most importantly, to provide the glimpse of the future that persuades you that this journey should be fuelled less by anxiety and more by excitement, ambition, and hope.

Part III – 'Day 2': The Reapproach

The Reapproach is where we set out on our adventure, it's where we activate our plans to step into the transition zone.

We'll meet the businesses and leaders who started their enterprises in Day 2, those who are journeying and exploring the uncharted waters, and importantly those who are navigating a pivot. In the new era of Day 2 we show how value chain patterns of Day 1 can be refined to be truer to the way in which we not only describe but also create value. We introduce the concept of the Total Value System to show how assets can drive financial performance over the longer term and create new sources of value that address pressing externalities and planetary and societal risks and needs.

We'll examine the 'art of the possible' and discover opportunity, collaboration, profit, and purpose in the process. We are on this journey together, and we want to share our way to get there, via 'Sustainable Performance'.

Sustainable Performance

We hope to take you to a place that is beyond compliance, that will call on the free market attributes of capitalism but not be enslaved by them. We are talking about a model that mitigates against the exploitation in the current economic model and that realises the opportunity in the ecosystem of the entire value chain. These ingredients will unite behind a meaningful and authentic purpose that releases energy and productivity amongst the stakeholders, and on the other side of this enhanced stakeholder performance will be superior financial performance. We call the performance delivered through this new operating system 'Sustainable Performance'.

Sustainable Performance is a way of operating that unlocks value creation advantages by placing the organisation in the service of an environmentally sustainable and socially just future. We'll demonstrate that many of the solutions needed to realise the opportunities presented are already in plain sight, and that business is perhaps the only institution that is powerful and agile enough to effect the change that will be required at this decisive moment in history, in this liminal space.

We are also believers in the power of storytelling. The Day 1 experience of paralysis in the face of the charging of the elephant contrasted with the Day 2 adventure of reapproaching and canoeing underneath these mighty creatures. This is intended as a useful narrative you can follow as we guide you through this unique moment in history.

References

Fink, L. (2022) Larry Fink's 2022 letter to CEOs: The power of capitalism. BlackRock. Retrieved 6 December 2022, from https://www.blackrock.com/corporate/investor-relations/larry-fink-ceo-letter

Fortune. (2019) Retrieved 6 December 2022, from https://fortune.com/fortune500/2019/

PwC. (2021) *State of Climate Tech 2021*. Retrieved 6 December 2022, from https://www.pwc.com/gx/en/sustainability/publications/assets/pwc-state-of-climate-tech-report.pdf

Rohr, R. (2004) *Adam's Return*, Crossroad Publishing, pp. 135–139.

Sayers, M. (2022) *A Non-Anxious Presence: How a Changing and Complex World Will Create a Remnant of Renewed Christian Leaders*, Moody Publishers.

Weybrecht, G. (17 January 2022) Business schools need to get serious about sustainability. *Financial Times*. Retrieved 6 December 2022, from https://www.ft.com/content/dc056f5f-2744-485e-a67f-362418c9375f

World Economic Forum. (2022) *Global Risks Report 2022*, 17th edition. Retrieved 6 December 2022, from https://www3.weforum.org/docs/WEF_The_Global_Risks_Report_2022.pdf

Part I

'Day 1': The Charge

Which enterprises will survive the reality of the climate crisis and transform through the transitional zone into Day 2? The drivers for success in business now need to respond to the deadline of climate change. Chasing an increasingly punitive cost of compliance may buy you time, or another day to face the ferocity of another charge. As Friedrich Glauner says:

> *Compliance is not the route to sustainable business conduct; the usual perspective on compliance is short-sighted and misses the point, because it rests on wishful thinking about how corporations should act and ignores the insight that it is the deep-rooted values forming a corporate culture which determine individual and corporate action, not legal decrees or voluntary self-commitments.*
>
> Glauner, 2017, p. 126

Unlike other business disruptions where we can find ways back to business as usual, this is not an option. We now need new knowledge, wisdom, purpose, and values to navigate the transition zone to the new era of Day 2.

Part I highlights the reality of this moment and the drivers of the transformation. In response, we explore the speed and boldness of the requisite response to avert catastrophic destruction. And in this place of uncertainty, we find opportunity drivers that could promise us the most fertile and exciting time that has ever existed for humanity to leap to a whole new way of life.

Reference

Glauner, F. (2017) Compliance, global ethos and corporate wisdom: Values strategies as an increasingly critical competitive advantage. In J.D. Rendtorff (ed.), *Perspectives on Philosophy of Management and Business Ethics*, Springer.

1

Reality

Today, Tomorrow, and the Peril of Business as Usual

The world is changing. We are in a liminal space and, as much as we might try to resist it, business as usual is no longer an option for the long term. The social and economic foundations that we have come to rely upon are being undermined and nature is becoming recognised as an unpredictable yet 'essential partner', if we are to continue to thrive.

At the same time, however, this 'essential partner' is showing signs of stress and even collapse as years of exploitation take their toll on fragile ecosystem balances we assumed would be ever unchanging. Due in part to these seismic changes, the external landscape affecting businesses is shifting, bringing with it questions without clear answers and a lot of uncertainty. In short, we have come to the end of the current era, and change now is inevitable, not optional.

The Size of the Challenge

Prior to 1946, when a group of surrendered German scientists strapped a camera to a V-2 ballistic missile, we hadn't yet seen the Earth from space. It was a fuzzy first, a black and white image, and remarkable, but it isn't the one most of us think about. We tend to remember the incredible globe shot, from the Apollo 8 mission, which was taken almost 20 years later, captured by Bill Anders as Frank Borman came round the moon for the fourth time.

Known as 'The Earth Rising' it changed the way we view ourselves forever; it's what Ron Garan, an astronaut from more recent times, describes as 'the orbital perspective' and what others have called the overview effect. Those lucky to see it for real from space often describe the moment as one when your awareness shifts to a profound and deeply felt understanding, that the Earth is a single organism floating in nothing and that all our actions as individuals are acutely and intimately connected and have collective consequences we should all take responsibility for.

The reason why this is important is because 'whole Earth thinking' or systemic understanding was not affirmed in the minds of leaders when they made their business and organisational decisions in 1760. From then until very recently, decisions were made through a mechanistic lens: establishing or creating a need, designing a product or service, extracting resources to create, before sale and distribution. From spices, to coffee, to cars, to computers; from entertainment, to finance, to app stores and virtual art – everything we have consumed since 1760 has had an impact on our Earth system.

Throughout the Industrial Revolution to recent times, business and government were working under the illusion of an infinite planet. There was an endless cornucopia of resources to uncover and transform into products, and along with them the services to get them made and sold. People built fortunes, and economies thrived.

The results from this rapacious and highly successful industrial momentum have been variously brilliant, uplifting, destructive, and depleting. However, as we haven't really created that many rules or guardrails for this growth, we now live inside the consequences of that uncontrol. With every extraction comes an externality, and externalities have piled up around us, in heaps of slag or plastic waste, in measurable molecules throughout our water and air, in social realities of cultural decline and inequity.

The impacts grew as the economy grew, through waves of feast and famine, literally and metaphorically, and as growth continued apace, the consequences of our collective actions increased alongside it. Year-by-year, the unmeasured negative outputs from the industrial engine at the heart of the Western drive for growth increased to the point of visible and measurable change. But in the last century, people started to question whether the endless growth was really sensible and as we became more connected and globalised the impacts became more apparent.

What started as a few powerful, pioneering voices at the beginning of the second part of the twentieth century turned into a 'green' movement and then a sustainable one. As scientific improvement and the computing revolution dovetailed into incredible capability, the truth of the industrial experiment became ever more apparent.

That truth is now unarguable – even in a post-truth world.

Every day, we have at our fingertips trillions of data points. They crunch through machine-learning algorithms, fed by off-planet satellites and on-planet sensors, all analysed by some of the smartest scientists we have. Together they represent a movement of understanding that means we now know what we are doing, that it is affecting the whole Earth system, and with a level of detail unfathomable in 1760.

So, we now know what's going on, we have the knowledge but not yet the action. The 'why' we aren't making decisions based on this understanding will be unpacked in other chapters, but the driver for the transition is clear: the whole Earth system has been pushed beyond its limits by an old era industrial revolution that has reached its natural end point.

What comes next, and is already transitioning, is the move to a fully sustainable system as part of the Fourth Industrial Revolution. The first three transitioned through different base energy supplies – coal to oil and gas, then to nuclear, and now in the fourth we have renewables. Along with the rise of our technological capability we have all we need to become fully sustainable, but as with all change, the incumbent and existing powers are struggling to release their grip, with all the impact that has on our world.

The Speed of the Change

Alongside the *size* of the challenge is the need to act at a speed in line with the challenges; that means acting now and learning how to move fast. Just as it is difficult for humans to comprehend the size of the whole Earth challenge, it is equally difficult to understand the speed at which the whole Earth is changing. In just 262 years we have, as a species, taken our world from one of relative balance, into one of rapid extinction.

Let's delve into that for a moment – try to stop and really think about what fast actually means in this context. Unless you are Marty McFly from the 1980s *Back to the Future* movies, when we think of time, we base our understanding on *our* experience of what time is – this means a human perspective, grounded in our 80 years or so, if we're lucky.

It is almost impossible to step out of this into 'longer time', where we try to wrap our heads around how things like entire forest systems, oceans, and atmospheres are *supposed* to change.

These are huge complex webs with multiple forces acting upon them, that *should* heat, cool, move, and change over long spans of time so that the living organisms within them can adapt, to evolve, and to survive. But if those changes happen so fast that we don't have that time, then the reverse of those three things happens: we don't adapt, we don't evolve, and we don't survive.

In the last 30 years, we have pushed our Earth survival systems into that space. We should be adapting, evolving, and learning to survive (and thrive) at the speed the changes are happening. But we aren't.

In fact, when you step back and look at just those 262 years, it's only in the last few decades of that time that the size of the impact has become clear, which has allowed us to understand the speed. And in response to that, we might as well be standing still.

This is not meant to create an emotional response of fear and inaction in the face of the danger, but there is no point in talking about drivers without them containing the truth.

As leaders, many have moved through the ranks of businesses or started them up and grown to size *inside* the principles of endless growth: profit at all costs, destructive competition, an irresponsible approach to externalities, and a myopic focus. The drivers for success in business did not contain sustainability, and it is only just starting to influence core decision making. This means the huge behemoths are hard to change and the humans that have benefited have very little motivation to do so. In their situation, why would change seem so pressing and why would they feel the need to move at speed?

The truth is they haven't. So, nothing has moved fast. There were PR-creating reduction strategies, covering the reduction of waste, water use, and carbon, but only at the tinkering edge. There is a reason that, despite the sustainability movement being around for a few decades, all the indicators of change have still been going in the wrong direction. The necessary speed of solutions has yet to be activated.

The other truth we need to face as leaders is that everyone has to act and now. The speed of change has outstripped our ability to adapt alongside it. We failed to act as a collective. Whether by governments creating policy levers, businesses making bold choices, consumers demanding change; none of these have happened at the speed we need. As a result, we are now left with a world system – the Earth model, the economy, and us – all wrapped up in an extinction-level speed shift without the required response.

Given we now have the data and many of the solutions to these problems, the challenge is how do we catch up? How do we accelerate the changes required? How do we all work together – government, citizens, and business? And, most pressingly, how do leaders make that actually happen?

Individual Versus Collective

Where we are is deeply complex and complicated, as the negative sustainability compounded impacts have become bigger than individual responsibility can ever solve. So, we need to address the conundrum of collective versus individual. We are still making decisions as though an individualistic mindset will solve systemic challenges.

It won't matter if, as a collective, only a handful of leaders individually build the best sustainability businesses. We need as many leaders as possible pushing as hard as they can to make changes equal to the need. There will, of course, be pioneers and pathfinders, but we must also find ways to lead together, to ensure as much change as possible becomes activated fast: collective, collaborative, and pre-competitive.

These are the principles we need to continue to embed in our decision-making processes, for as the negative drivers for change continue to expand from rivers to oceans and city air to nation sky, we must also expand, from linear supply chains to systemic value chains, leadership by winning to leading through lifting all.

Topics such as whole system climate response, or systemic protection of entire oceans or rainforests, are deeply challenging for our current organisational model of individual competing nation states and singular organisations and businesses. However, that's where the drivers are pushing us; to make decisions that connect to the collective and as leaders we must respond.

In recent history, we have seen collaborative groups attempt to make change happen, whether at a United Nations and governmental level, such as the annual COP meetings (27 and counting), or meta-platforms of targets and commitments, such as the Sustainable Development Goals and the Net Zero movement.

But despite the science and visibility of the challenges becoming irrefutable and fully transparent, the responsibility to tackle them in a truly connected way remains a hot potato thrown from government to business to consumer to citizen, with only a few seeming to take direct leadership or responsibility in the way we actually need.

We *have* seen some very encouraging indications of travel, especially from the more enlightened businesses and governments, but so far the outcomes of these endeavours have yet to equal the impacts they aim to reverse or balance. It is vital that we don't get disheartened and disaffected by the lack of movement, but instead try to believe that these growing bodies for transformation will hasten their efforts in light of the increasing need.

Working across a spectrum of businesses, we see a big shift beginning to happen that is useful as an indication of hope. Although there is still a lot of competitive focused action, 'how can we be the best sustainability company?' – and that can be useful to drive change – we are also seeing a real acceptance of how business can start to tackle the connected challenges.

It will be in the systemic solutions that we find the real levers for movement. Lots of different businesses use the same mines, factories, and distribution and sales channels. As they begin to collaborate more effectively, we will start to see the transformations happen.

Some of this is already at the edge of visibility. If we look at the emergence of new, circular materials in packaging or textiles we can see the start of entire new business models emerging. These provide companies with novel ways to delight consumers and attempt to maintain or grow sales. If you combine this with natural capital accounting, whole system change may well be possible.

These glimpses into the possibility of systems change show a deepening awareness, acceptance, concern, and energy towards the environment and climate crisis, its resultant societal impacts, and the risks of not making big changes. As part of this shift businesses are making pledges, setting goals, and seeking to make choices in line with a more sustainable business and world.

So, what has changed?

In 2014, French weather forecaster Évelyn Dhéliat predicted that by 2050, temperatures in France could reach up to 43 °C in August due to the climate crisis. Her predictive weather report was a sea of intense orange and red and she stated that this was what France faced if we didn't take action to tackle the climate crisis.

In June 2022, her forecast came true, 28 years earlier than predicted. A few years ago, we might have likened the climate crisis to an elephant that periodically appears, threatens to charge, looks intimidating, and elicits fear, but then retreats for long periods of time. Increasingly we are coming to realise that the elephant no longer retreats. It is facing us down, waving its head and ears, and increasingly 'running at us at full speed'. Climate change is not something happening in the future. It is here now, and we can't ignore its warnings.

Climate-related events and disasters are increasing in frequency and intensity. Record temperatures, flooding, wildfires, and unprecedented climatic events are reported weekly. The year 2022 alone saw catastrophic floods following an unprecedented monsoon season in Pakistan, and concurrently a 'heat dome', causing prolonged extremes of heat, settled across Europe, North Africa, the Middle East, and Asia. Temperatures exceeded 50 °C (Iran recorded over 53 °C on 5 August 2022), many records were broken and in places unused to such temperature extremes, such as the UK, temperatures topped 40 °C, leading to anomalous wildfires, droughts, and water shortages.

The still-rumbling global Covid-19 pandemic provided a backdrop to these crises and numerous others preceding them. The death toll to date exceeds 6.5 million. This devastating virus swept across the world, ignored borders and exposed our vulnerability to naturally occurring pathogens, exacerbated by human-induced environmental destruction. We were forced into a global hiatus (acknowledging the bravery and sacrifice of essential service providers). As we took a breath and acknowledged our ability to truly control nothing, scientists became our guides. They brought glimpses of clarity in a time of deep complexity and garnered a respect for their expertise, which hinted at safety in a very unsafe place. Perhaps this response triggered a similar increase in trust for the climate science that epidemiology is inevitably connected with. Perhaps too, Covid-19 demonstrated that a crisis at a global level can happen, and quickly. Or maybe the lockdowns, restrictions on our lives, and total disruption to everything that was once 'normal' taught us that what seemed unimaginable before the crisis, was entirely predictable when we were in the midst of it.

A global pandemic also illuminated the unacceptable chasm of inequality woven into the fragile threads of our global society that hold us together. The climate crisis and the disaster and degradation it brings similarly exposes this inequality. The perceived recent awakening and enlightenment towards this looming climate crisis is largely a developed world phenomenon. Around the world, particularly in less developed regions where millions live at the margins of environmental vulnerability, people have experienced the acute impacts of climate change for many years, and many have been furiously sounding the alarm – people and places that have less political power and influence to make the right decisions at a global level.

Friends of the Earth tell us that climate justice 'recognises the climate crisis as a social and political problem, as well as an environmental one'. It recognises that the effects of climate change will disproportionately impact those who are most vulnerable and marginalised. Worldwide, for example, despite bearing the burden of multiple household and livelihood responsibilities, 'women have less access than men to resources such as land, credit, agricultural inputs, decision-making structures, technology, training and extension services that would enhance their capacity to adapt to climate change' (Osman-Elasha, 2012). The inequalities that mark the social, economic, and cultural fabric of our world also mark the crisis of climate we are facing. In fact, it could be argued that injustice is a root cause of the climate crisis, or at least our ability to effectively respond to it.

This book focuses on business – and we refer largely to companies that are based in the global north – but of course in our globalised world many businesses have operations and supply chains in less developed nations, where the vulnerability and exposure to climate change is most evident. Climate impact and social inequity are two sides of the same coin. Businesses that are to thrive in our transforming world will need to understand and respond to the full context of their business ecosystem, including in areas where climate impacts will be most acutely felt.

Pivot or Perish

So, momentum is building. The trend among businesses, cities, and governments is to set carbon reduction targets as the embarkation point to the climate transition. The shaking loose from the old era is happening, with all the potential chaos that will arise as global forces and philosophies collide.

Set in this emerging context, what we really want to talk about in this book is how you need to embark upon a 'pivot and perform' journey, but before we get there, we should call out the risk of a 'pivot to buy time and perish' approach. What do we mean by this? We are arguing that such is the significance of this moment and the changes that will be thrust upon businesses, that those businesses that refuse or are unable to greatly adjust their ways of working are ultimately likely to perish. Those that see this agenda just as a cost of compliance issue will buy some time and earn their licence to operate, but ultimately this will not be enough. To 'pivot and perform' will require something much more transformational.

Environmental, Social, Governance – ESG

ESG is a challenging topic. For something so significant, it's remarkable that it's so unclear and so challenging to define. Born out of sustainability reporting, the understanding and use of it ranges wildly and widely. Although the impact of ESG reporting varies at the investor or company level, at a market level its impact has been dramatic, and for the most part positive.

As we shall explore further in subsequent chapters, ESG has added a much needed framework to help companies move towards better sustainability, even if it's still so difficult to agree on. Some use it as the baseline for redesigning their business, others as a way to do the minimum and keep investors happy. But most importantly, it is increasingly used to make decisions about where the money goes and how it flows through a business and that's why it's now the biggest driver of change.

At its most hopeful, ESG tools and analysis feel like the starting point of a journey that will end with a sensible and practical foundation for what all companies and organisations should have as a sustainability minimum, with options to scale up to much, much more. A comprehensive and transparent way to see who is playing the game and how hard they are playing, or a guide for internal improvement and external judgement. Alternatively ESG is seen to be an interference to the purely fiscal ways of evaluating an asset, and it will lose the battle for influence and relevance. Perhaps, and most likely, it will become a compliance tool against which assets will be scored and a further data point to allow for more considered deployment of capital. In Chapter 4 we shall explore the way sustainability is described, and often misunderstood, by different sectors, but ESG is the terminology that is gaining traction amongst business and finance leadership. For example, internet searches for the term ESG have grown five-fold since 2019 and money flowing into sustainable funds increased from US$5 billion in 2018 to nearly US$70 billion in 2021 (McKinsey, 2022).

For Gina Mastantuono, CFO of ServiceNow and a member of the Forbes Next 50 2022 list, this growth presents both a challenge and an opportunity.

> *There is a clear appetite from investors for information on ESG performance. ESG goals, activities and reporting must be embedded into business processes and policies to succeed. CFOs need to know how to respond and lead in this emerging landscape. The good news is that purpose and profits are not mutually exclusive. ESG's ability to create long-term value makes it so we don't need to sacrifice in the face of short-term economic headwinds. But we've got to get this reporting right.*
>
> Mastantuono, 2022

For ServiceNow, unpacking the complexity implicit in ESG data, disclosures, and reporting, 'leveraging governance to bring efficiency and accuracy into this process', presents both a business opportunity and a way for them to 'help in the transition to a net-zero future'. Their newly launched ESG Command Centre solution, for example, will bring together ESG goals and metrics in one place and help to demystify the numbers (Mastantuono, 2022).

Mastantuono clearly sees both need and opportunity in the agenda, but for others ESG presents challenge and disruption.

In a recent conversation, a director for innovation told us that they had to 'change their entire investment strategy to align with sustainability because of the whole ESG thing'. With trillions of dollars under management, the potential impact of ESG on markets is remarkable. Despite its nebulosity, ESG has become a massive driver of the sustainability shift. And yet, the uncomfortable truth remains that we need to move beyond the ESG reporting and compliance for ESG action to have a meaningful impact on reducing carbon emissions.

Whatever your view of ESG as it stands today, businesses and organisations will be increasingly scrutinised from now on as a continuum of how they perform against the needs of the environment and how they perform against their treatment of people. This is not going away. We doubt there will be a return to 'business makes profit at the expense of. . .', no matter how many articles about woke capitalism are written.

How Burnt Is the Burning Platform?

We argue that limited change to date is the consequence of the two dimensions we outlined at the beginning of this chapter – the size and scale of the challenge and the speed at which change is occurring.

Size of the Charge

The foundations and strongholds upon which the businesses and the institutions that govern them are built are not designed to confront the sheer size of the Charge we are facing. As we have said, we are in a liminal space, our old ways of living, working, and using the world's resources are no longer sustainable. Like it or not, we are being forced into a new era that holds deep uncertainty alongside incredible opportunity. But rather than seeking to pivot, to explore new opportunities that will help to solve the huge problems we face, the systems of what is now the old era are doubling down. They are doing what they know best to confront an unknown foe and hoping it will work. But incremental shifts and changes take us nowhere near the transformation required. Businesses need to pivot in this moment, or ultimately, they will perish.

We are not suggesting that regulation isn't essential, but it isn't enough, and arguably the noise and confusion it creates for businesses and leaders is leading to stagnation, frustration, and even fear, rather than inspiring the action, innovation, and enterprise required for the new era.

But we are seeking hope, not panic and as we will explore, there is opportunity, calm, and relative safety to be found in the Day 2 experience of canoeing under elephants.

Urgency

Sustainability is our business; we face the 'reality of now' every day, we know the gritty details. For many the sheer urgency of the crisis has not yet sunk in. And yet one thing that inspires innovation, creativity, and action like no other is urgency. In the face of an imminent threat, with the inevitability of stepping into the danger zone, we find incredible new ways to get from A to B. In the face of the Covid-19 'Charge', normal barriers to action dissolved and the power of enterprise was released to bring about solutions.

The Ventilator Project

We're sure that you remember the uncertainty and fear at the start of the Covid-19 pandemic. In those initial months, no-one really knew what to expect or how best to prepare for what was to come. However, one thing became clear very quickly to those living in Britain: the UK was lacking the volume of ventilators its hospitals would need. That was when the government turned to industry for help.

Adam Blomerley, CEO, and Ian Quest, Director, of the data and consulting business Quick Release were part of a consortium, which included Ford, McLaren F1, Airbus, and other data and engineering companies and consultancies, that came together to respond to a call for support.

> *The government put out a challenge: whoever could make hospital-grade ventilators within certain time frames would be able to sell them to the UK government. Hundreds of people put their hands up and said, 'Yeah, we can do it.' – from people in their sheds to large global manufacturers. It took around 100 days to go through design, tests, certification, set up the manufacturing, set up the testing, set up the supply chain, and then build 12 000 of them.*
>
> *For the people involved, it was a compelling and emotional journey to be on. It was all about the art of the possible, rather than focusing on what we could not do. We knew that every day 1000 people were dying and that was a strong motivating factor.*
>
> Blomerley and Quest, 2022

A compelling purpose unleashed enormous amounts of energy and this story serves as a reminder of the productivity you can get when you engage with people at an emotional level. It also shows that when there's a burning platform, companies quickly come together in a pre-competitive collaboration space. Normal data privacy and competition rules are pushed to one side. In the ventilator project, the reality of thousands of people dying took precedence over any historical barriers that are typically in place between organisations when you are outside a crisis.

Returning then to our concern – pivot, buy time, perish. There will be many businesses that will become consumed or paralysed by the Charge. Others will believe that responding to incremental compliance nudges will allow them to return to business as usual (in the safety of their stronghold). The risk we would call out is that if you survive the Charge, but fail to leave the bush, you'll only face the Charge again. However, by surviving the Charge you've bought yourself a licence to get in the canoe – and you need to get in the canoe if you want to find performance and value in the new era.

While perish may seem a strong word, and perhaps a gradual disintegration may describe the likely outcome more accurately, we need performance and prosperity to be the outcomes. Compliance won't engage with the changing customer demographic you need to appeal to. It won't deliver the authenticity and ambition that you are likely to need to attract the best talent. It won't make the audacious strategic decisions that will be needed in aligning to the markets of the future. It won't move you from a supplier leverage model to one of supplier loyalty and it won't solve the global existential crisis we now face.

We are not suggesting that compliance should be neglected. If Stuart and Louisa hadn't complied with the guide's instructions, then they would likely have been trampled. But Day 1 is about survival so that you get to live out the adventure of Day 2.

References

Blomerley, A. and Quest, I. (13 July 2022) Interview with the authors. Friends of the Earth, www.friendsoftheearth.uk

Mastantuono, G. (14 September 2022) Interview with the authors.

McKinsey (2022) ESG is essential for companies to maintain their social license. Retrieved 7 December 2022, from https:// www.mckinsey.com/capabilities/sustainability/our-insights/ does-esg-really-matter-and-why

Osman-Elasha, B. (2012) 'Women…In the shadow of climate change', *UN Chronicles*. Retrieved 7 December 2022, from https://www .un.org/en/chronicle/article/womenin-shadow-climate-change

2

Strongholds
The Fantasy of the Status Quo

In this chapter we will look more deeply into the idea of strongholds and set out how the reality of now and the scale and urgency of the disruptions brought about by climate change and associated phenomena undermine them. We'll look at how the dismantling of old strongholds of the current and passing era will bring anxiety in place of security. And while some will try to pretend that the nature and scale of change is not that serious, presenting a fantasy tale in place of the truth, for the wiser, open-minded and pioneering leaders, there will be excitement in the opportunity. An opportunity to recreate a new form of stronghold that secures the future for the many and not just the few.

The Mountain is Broken

The people of Sicily's second largest city, Catania, need no reminding that their existence is linked to the temperament of their volcanic companion, Mount Etna. Over many centuries, the city has been used as a refuge against the unpredictable moods of its powerful neighbour. But it has been unable to escape its force. Visitors will quickly observe the many signs of episodic destruction evident in the modern-day expression of the city – each historical episode representing an opportunity to rebuild and to reimagine, providing a diversity of architectural styles and the aesthetics mirroring the cultural, societal, and economic influence of a moment in time.

But Etna doesn't just take. Much of the city, including the roads, are made with lava. The volcanic soils are some of the most productive and fertile in the region with vineyards and orchards occupying the lower slopes of the mountain. It is known for its rich biodiversity and as a tourist destination, including a modern ski resort. More recently it has received considerable interest in geothermal energy as a cheap and renewable energy source for local communities. The moods of the volcano are carefully monitored with sophisticated seismic technology providing an early warning system. As a consequence, communities have reinhabited the slopes of Etna, living confidently in partnership with the volcano and with nature.

This reassurance comes from reliable science and predictive modelling, which allows a different relationship with the volcano. But we are recent beneficiaries of technology that provides such anticipatory outcomes. Over the centuries, we can imagine the frequent cry of *Scassau a Muntagna!* (Sicilian dialect for 'the mountain is broken'), as Etna unleashed large and small assaults on the local communities – and with that cry, the fear that comes with uncertainty as to how merciful the mountain will be on that day.

Today there is a rhythm to the coexistence between Etna and local communities. The scientists do their work. They monitor, they communicate. The local inhabitants respond, they take what assets they can and, if necessary, they migrate to a different place.

By listening to scientific wisdom, to truth, we can avert destruction where, in the past, thousands have died and where communities and the city of Catania were almost completely destroyed. This source of historical terror has been reapproached.

The Intergovernmental Panel on Climate Change (IPCC) is the United Nations body for assessing the science related to climate change. The IPCC was created to provide policymakers with regular scientific assessments on climatic shifts, the implications, and potential future risks. As well as putting forward adaptation and mitigation options, it gathers leading scientific institutions around the world to assess climatic impact based on the latest science. It is through the IPCC that thousands of experts from around the world are able to synthesise the data points and form conclusions on the vulnerability of the earth's systems alongside mitigation and adaptation responses.

Since the formation of the IPCC in 1988 they have been warning the world of the impending crisis. In 2022 the message shifted. 'The scientific evidence is unequivocal: climate change is a threat to human wellbeing and the health of the planet' – if we are to avert the crisis, 'it's now or never'. The mountain is broken!

We have observed this ever more scientifically robust and clear statement being made by the IPCC over 30 years. And yet, with the 'glowing avalanche' of evidence, with magma flowing down the sides of the volcano towards billions of inhabitants, apathy rather than urgency has characterised the world's response. Why?

In this chapter we try and find the answer.

The Nature of Strongholds

In the introduction we said that understanding the nature and power of strongholds would be a theme that we want to explore in the book because in a liminal space, strongholds become threatened as everything is changing. Strongholds describe those social, economic, political, or institutional structures that have been built to retain power and control over storehouses of wealth. They are places of security for those who are within the stronghold and who abide by the rules. Creating strongholds is instinctive and

not inherently bad. It is in our nature to protect and pre-
serve, to store up and create places of safety. We do this
naturally in our family units and among communities, creating places of healthy refuge where we enjoy familiarity
and commonality. We should absolutely aspire to build or
find these places of security and stability.

However, strongholds can become unhealthy. They
can become vulnerable in their isolation as 'strongly
defended' walled fortresses if we ignore the signs.

We observe that many of the existing strongholds we
know and rely on are shaking and many will be destroyed.
We are in a liminal space, the space between spaces, at the
end of one era but before the start of the next. The next
industrial revolution is upon us, driven largely by the climate transition. This moment presents a rapidly changing
risk and opportunity landscape that needs navigating. But
the seismic shifts are being measured, we have foundational truths, and against the backdrop of unequivocal evidence we need to make some decisions.

This comes with responsibility. For many leaders, compliance represents an easier route. It's a level playing field –
you have to retain your licence to operate to maintain your
business as usual. Tell me what to do, legislate, and we will
respond, often at a pre-competitive industry level.

Compliance is necessary, but an overly myopic focus
results in leaders missing the opportunity for value creation,
for discovering the adventure of Day 2. Leaders who are
most likely to thrive in this space will need to be courageous
and willing to step into a place of uncertainty, possible danger, and adventure with the right guides alongside them.

Our message to leaders is clear. You have to move. The
liminal space that you find yourself in is real. Transformational
change is inevitable, and if you don't respond to the imminent

danger then the organisation that you lead will likely perish. For this, you will need to embrace the reality of adventure, and the likelihood of having to leave strongholds behind.

Typical business models and value chains reflect prevailing norms, often with a leadership focus that is self-serving and where strong values are left at the door of the stronghold. As a result, many businesses and business models have become stronghold fortresses that must be defended at all costs, and board directors, guided by audit and risk management controls, are encouraged to fortify legacy strongholds as their fiduciary duty.

Inside the stronghold is perceived to be secure. Anyone outside the stronghold may be welcomed if their involvement strengthens further the storehouses of the stronghold. Alternatively, those outside the stronghold may be viewed as a possible threat of attack. In this liminal space where everything feels uncertain and people have a lot to lose, many strongholds are closing in, shoring up their walls, and strengthening their defences. In many ways this is understandable. If building strongholds – and protecting and providing for those we care about – is instinctive, then you will naturally fight if your stronghold is threatened. We understand the complexity of this time, and as we travel through this story, we shall seek to acknowledge the challenge but propose a different way forward.

This chapter explores some of those strongholds and the sources of attack. There are too many strongholds to list here and we will all have strongholds that are personal to us. So the following are by way of example.

We also examine the emergence of the fantasy realm and how this is being embraced by some leaders as it represents an opportunity to move to a comfort 'off ramp', where we can convince ourselves that a threat is not real. We will signpost the risk of stepping out of truth into fantasy, which only serves to delay the inevitable and may make the destruction of strongholds more explosive. But before we get going, a word on capitalism.

Is Capitalism a Stronghold?

It's easy to shy away from subjects that typically polarise audiences. Capitalism is one of these. While we would suggest that capitalism is not a stronghold, modern versions of it have become strongholds. Perhaps more importantly the battle over the various forms of capitalism promoted in different geographies, cultures, and industry sectors are a competing force in the shaking and eventual destruction of strongholds.

Opinions range from those who believe capitalism to be the source of much of the environmental and social degradation that characterises the systemic breakdown we see in the world, to those who see it as the vehicle that has lifted millions of people out of poverty by releasing human endeavour to do extraordinary things. As Bono, lead singer of rock band U2, describes it: 'Capitalism has taken more people out of poverty than any other "ism". But it is a wild beast that, if not tamed, can chew up a lot of people along the way' (Stolworthy, 2019). Whether capitalism is your friend or foe, and however hard you challenge it, we see capitalism as an economic model that will transcend eras.

At its core, capitalism is a model that relies on market forces rather than control by a central government. Capitalism is an exchange mechanism for bartering goods and services in an open market of buyers and sellers. It is a system defined by free markets, competition, and the search for uncontested space. A system that is agile, innovative, and entrepreneurial, and a system that, in our opinion, is more likely to bring about the innovation and the leadership that will act as essential hinges to us if we are to navigate successfully between eras.

But we recognise how it can become a 'wild beast'. It has been proven to be malleable to the prevailing culture of the day, or of the nation, the city, sector, or company. Modern parlance describes it in many forms. An individualistic form of capitalism reliant on self-interest and promoting greed, or alternatively as a form of wealth generation and capital deployment for impact embraced by an inclusive or responsible intent. We hear a lot currently about 'woke capitalism', a term coined in 2015 by Ross Douthat when writing a piece for the *New York Times*. He defined it as the way that companies signal their support for progressive causes in order to maintain their influence in society. These progressive causes are noted to be typically from the left of the political spectrum. While the emergence of 'corporate wokeness' in large corporations may give a helpful hand to many of the subject areas of this book, we prefer to focus on the more inclusive form of stakeholder capitalism. Professor R. Edward Freeman is credited with developing the principles of stakeholder capitalism into a practical business model. He says, 'It's about how business actually works in the real world' (Feloni, 2021).

The concept of stakeholder capitalism is straightforward. Companies should make money and use this wealth creation to generate returns for shareholders, tackle inequality, and regenerate and protect our natural assets. They do this by engaging the stakeholders in the value chain to generate value and then to distribute this value across all the stakeholder groups that are relevant to the entire value chain of the business. As we ponder the potential opportunity landscape while sitting around the campfire after Day 1, we'll see how Day 2 is only possible when we integrate our value chain in its entirety and embrace stakeholders in this more inclusive and collaborative way.

Stakeholder capitalism offers some fundamental but subtle differences from the neo-classical models of capitalism that form the narrative of the form of capitalism that has become dominant in the old era. For example, the idea that corporations shouldn't carry a position of social responsibility and that only individuals or the public sector should carry this responsibility. You try telling that to the average CEO who is striving to navigate their corporate positioning around institutional racism, diversity, equity and inclusion (DEI), and climate change, with the consequences of getting it wrong including loss of talent, destruction of brand value, and being alienated from certain market and investment sectors.

Stakeholder capitalism is capitalism that can't be segregated from the environment and from society, that operates in a network and that gives recognition to the opportunities to make a positive impact in a wider ecosystem.

We resonate with the reflections of our interviewee, fashion leader Geoff van Sonsbeeck, the CEO of leading sustainable fashion brand House of Baukjen, who shares: 'I think that shareholder capitalism is pretty much over the way we know it. However, at the same time, stakeholder capitalism is just a long-term version of shareholder capitalism. There is no long-term shareholder capitalism without thinking very hard about your stakeholders. I think it's good business full stop to think outside the short term'.

Industry Strongholds

The largest and most visible climate impacting stronghold of the old era is the fossil fuel industry. A 2021 study in the journal Nature (Welsby et al., 2021) found that in order to avert the worst impacts of climate change, most of the world's known fossil fuel reserves must remain untapped. According to the study, 90% of coal and nearly 60% of oil and natural gas must be kept in the ground in order to maintain a 50% chance that global warming will not exceed 1.5 °C above pre-industrial levels.

Now, a new study (Chen et al., 2022) led by researchers at MIT estimates the global net present value of stranded assets in coal power generation alone through to 2050 ranges from US$1.3 to 2.3 trillion. Separately, in July 2022, the British newspaper The Guardian, reported that for the last 50 years fossil fuel companies have made US$3 billion per day (Carrington, 2022). These staggering numbers are based on World Bank data, verified by three expert institutions, and show just what is at stake both for this giant industry as its stronghold begins to crumble, and for the world and humanity if the industry resists.

Carbon Tracker is an independent financial think tank that carries out in-depth analysis on the impact of the energy transition on capital markets and the potential investment in high-cost, carbon-intensive fossil fuels. Its founder and Executive Chairman Mark Campanale, states: 'To keep to 1.5C, this means [international oil companies alone] forgoing around $100 trillion of potential revenues. You can see why oil oligarchs and nations controlled by political elites want to keep their fossil fuel rents, the source of their power' (quoted in Carrington, 2022). These are the reinforcing national and regional strongholds at work to protect their unimaginable wealth creation.

Maintaining global warming to 1.5 °C will require a 50% reduction in carbon emissions this decade, 50% next and net zero by 2050. This commitment, made by over 190 countries and now swathes of the world's largest corporations, is a commitment to dismantle the fossil fuel stronghold. Of course, we shouldn't just see big oil and gas as a stronghold. They have further reinforced the walls of national, regional, and other institutionalised strongholds with their unimaginable wealth creation.

Within the fossil fuel industry, facing enforced transparency and an unfolding of the reality of both their wealth and impact, CEOs are having to find a clear path. Major oil companies such as Shell have committed to a net-zero target by 2050 with 50% reductions in their own emissions by 2030. The debate rages about the integrity of these commitments and the validity of the numbers that underpin their claims. Professor Paul Ekins at University College London is sceptical, and referencing The Guardian's claims, says, 'At the very least these companies should be investing a far greater share of their profits in moving to low-carbon energy than is currently the case. Until they do so their claims of being part of the low-carbon energy transition are among the most egregious examples of greenwashing' (quoted in Carrington, 2022).

If you are the CEO of a large polluting industry, you are likely to be surrounded by a lot of 'shouty' voices. Your employees, customers, your board, and investors will all have contrasting demands. Some will demand that you just comply, tick the boxes, and get back to doing what we know within the model of the last 80 years. Others will state it doesn't matter how firmly we build the walls around our existing model, it will not be able to withstand the ever-increasing power of the elephant charge. At the same time national governments and supranational trading unions are increasingly seeking Day 1 compliance. The introduction of climate risk related legislation across the world is one of the 'compliance sticks' that is making the Charge more powerful, more threatening, and more costly. For these industries, the ferocity of Day 1 is becoming more evident, and the strongholds are shaking.

Financial Strongholds

Increasing legal compliance isn't the only thing shaking this stronghold. Let's turn our attention to the finance sector and the influence of investors. One of the largest institutional investors in the world, AXA Investment Managers, states:

> *Investing in companies and projects that are leading the way to a more sustainable world not only helps us reach a net-zero economy by 2050 but can deliver more sustainable returns in the future. The vast majority of our assets under management integrate our ESG analysis and quantitative scores into the investment process, while applying our core exclusions policy. We believe this can deliver value for clients by identifying risks and opportunities linked to key sustainability trends in the global economy.*
> AXA Investment Managers, 2022

In this one paragraph, AXA Investment Managers refer to the Day 1 compliance experience of ESG analysis and quantitative scores, but then hint at the 'risks and opportunities' of a Day 2 adventure, with the belief that this is an agenda that will deliver value.

AXA is considering their own position on this agenda but then extend their influence to a much wider target audience. In an act of mutual Day 1 enforcement, AXA have introduced a policy claiming: 'We have reinforced our climate engagement and stewardship policy, with a view to divest from climate laggards (a focused list of companies considered material in both portfolios and on climate change) if after three years of constructive engagement not enough progress has been made.'

The context within which institutional investors such as AXA are operating is changing. Upstream and downstream in the value chain of investors we are seeing a more influential and holistic ecosystem at work. In the UK, the Make My Money Matter campaign is on the case of institutional investors. They claim to have seen huge progress in aligning UK pensions to net zero. By January 2022, £1 trillion of UK pension money is now in programmes committed to robust net-zero targets, with a halving of emissions before 2030.

Then we have the influence of the credit agencies. Richard Mattison is President of Sustainable 1 at S&P Global. This unit was set up in April 2021, on Earth Day, and acts as a dedicated focal point for all of the sustainability efforts across S&P Global. This company is the world's largest rating agency and has many major investment benchmarks around the world, including the S&P Dow Jones indices that runs the S&P 500.

In conversation with Mattison, he highlighted the need for a vast increase in the allocation of financial investment towards sustainability amounting to 'trillions of dollars' to ensure that global sustainable development and climate goals are to be met, particularly in developing markets. For Mattison the narrative 'is about growth, as well as mitigating risk', something which is often missed. S&P plays into this narrative by providing 'objective and independent intelligence to allow investors and banks and other allocators of capital to make better [and more effective] decisions', to better understand 'the risks associated with any particular investment'. Critically this knowledge is now leading to action. For instance, 'in the year to date [October 2022] sustainability has impacted close to 10% of corporate and infrastructure credit ratings actions. Note the impact can be positive or negative or affect the outlook'. This is how ESG is starting to bite in the investment markets.

Inheritance and legacy amongst business leaders play into expectations on action. While writing this book we had a conversation with a client who is using the G in ESG as a way to change the way their top executives are remunerated. With a 2030 set of sustainability targets, the reality of measurable carbon reduction is about to become a key factor in how they get paid. Pressure is being applied and budgets provided to go and find the carbon in the business and reduce and remove it.

In September 2022, Thomson Reuters reported on the latest US Securities and Exchange Commission findings. 'The securities market regulator has various ESG-related rulemaking projects as well as an executive compensation rule that was finalised late August 2022' (Ho, 2022). The Thomson Reuters article cites a May 2022 report by consulting firm SpencerStuart, where companies surveyed were increasingly incorporating ESG goals and metrics into many elements of business: 71% were incorporating ESG goals and metrics into overall company strategy.

Another 52% into integrated risk management, 48% into criteria for director appointments and 46% in executive compensation.

The message is clear. As the CEO of a large financial institution told us recently: 'I now accept that I can't keep throwing money at my CSR department and telling them to take this issue away from me. I need to lead a change programme from the top.'

Consumerism Strongholds

Consumerism, our insatiable desire to acquire more, is a stronghold that originates in the early twentieth century. But the walls of consumerism have been reinforced and roots deepened ever since, by experts who know how to manipulate our appetite for contentment. But consumerism is a stronghold that relies on the infinite supply of raw materials from a single planet with finite boundaries. The fundamental weakness in the longevity of this stronghold is clear. Layer upon this a shift from companies having zero responsibility for the product after point of sale to the increasing demand, both from consumers and regulators, for producers to retain responsibility for the retrieval and re-stewardship of their product at the end of its life. With the disruption caused by a producer-centric responsibility amongst our customer base, we can imagine why the desire persists for this stronghold to remain like a fortress. After all, many businesses depend on people desiring things, 'needing' things and buying things. In the future businesses will also depend on people returning things.

Businesses are experts in forward logistics; they make and create things and sell them on. But as the reality dawns that this is not a model that can be sustained in a finite world, pressure builds for these companies to take more responsibility. As the stronghold begins to shake, they will need to consider their logistics in reverse and take greater responsibility for their products and waste. It's a huge task that many are resisting. As the need grows to embrace stakeholders across the value chain, so the alternative pathways to Sustainable Performance and value creation in the new era become more focused.

Obsession with Conformity Strongholds

Conformity appears to be part of human behaviour as does the aggressive response received by those embracing non-conformity. Social pressure leading to a change in social expectations takes time. We observe increasing social division and tribalisation around issues and beliefs that influence behaviours often in echo chambers where people just want to fit in. Following the tribal masses, even if it's into a fantasy zone, can feel more comfortable than non-conformity in the truth realm.

This is, arguably, the greatest stronghold on professional behaviour and norms that will hold people fast in Day 1. Conformity is a real blocker to adaptability and the personal freedom to let go of past 'badges' of professional approval. Often enforced by the isomorphic rigidity of résumés that enforce professional conformity along set pathways, these strongholds come in many societal norms and forms. We can easily be transfixed by Day 1 competitors and institutions rather than stepping into the liminal space of new venturing and new pathways of differentiated purpose, organisational belief, and reason for being. The Day

2 native canoers we describe in Chapter 7 have not conformed to prevailing industry norms so that they can perform in the new era that is dawning. They also recognise that non-conformance doesn't win awards until later adopters join them.

Another manifestation of conformity at work is the short-term reporting cycle of public companies. In 2009, on his first day as CEO of Unilever, Paul Polman said to investors:

> *Unilever has been around for 100-plus years. We want to be around for several hundred more. . . So if you buy into this long-term value-creation model, which is equitable, which is shared, which is sustainable, then come and invest with us. If you don't buy into this, I respect you as a human being, but don't put your money in our company.*
>
> Skapinker and Daneshkhu, 2016

In a bold and potentially risky move, Polman was calling out the contradiction of short-term investor horizons and the need to build long-term value. In 2009, Polman was seen as a maverick; now there is widespread acceptance that quarterly reporting leads to myopia in business executives.

ESG and the concept of Sustainable Performance are longer-term aspirations. The stronghold predicated on the promise of a quick buck is being shaken.

Earlier we called out the habits of leaders who have tended to prevail in the era we are moving away from. But we operate in a construct of doing business that has been embedded and amplified by the MBAs, the business textbooks, the established set of business and accounting rules, and business school teachings. The holding on to old era formulas inevitably wire leaders for old era outcomes. As Henri-Claude de Bettignies, Emeritus Professor of Leadership and Responsibility at INSEAD, says:

> *Leaders, in all walks of life, should see their contribution in making the future better as a categorical imperative embedded in the very nature of their position. Moral leadership thus becomes a prerequisite ingredient in any contemporary effort to further sustainable development. Today, in a world which is confronted by so many fast-changing and difficult issues, moral leadership is indispensable to induce change and in a direction that will further the common good. Leadership education needs to be infused with case challenges that call for just and responsible decision-making beyond abstractions and models of leadership.*
>
> De Bettignies, 2017

Mindsets can become entrenched in leaders as a consequence of the conditioning they receive in business school education. In conversation with Henri-Claude he goes on to say: 'I am critical of the way business schools are preparing people who have so much power. That's the power of responsibility not only for customers but also society'.

Dame Polly Courtice of the Cambridge Institute for Sustainability Leadership (CISL) says: 'We only exist because so many business schools aren't doing their job properly. The fact that CISL continues to be so busy says a lot about where the business schools are' (Courtice, 2022).

We recognise that there has been a growing and significant movement to bring greater impetus in research and teaching for many business schools, notably through signatory membership and reporting under the Principles for Responsible Management Education (PRME), a United

Nations-supported initiative founded in 2007. PRME's vision is to create a global movement and drive thought leadership on responsible management education, but the old era stronghold in business academia restrains the transformative wishes of many business faculties.

Despite the deep complexity that future leaders will be required to navigate, the new generation of leaders are still fed the language of the business stronghold of the old era. As a business leader, you are less likely to be criticised for following perceived old era wisdom. You keep your record clean and work the hierarchy, setting career stardom ahead of personal values. You may become blind to common sense, and entrepreneurial opportunities. Instead, you manage the stronghold, and if you succeed according to the rules of the stronghold then at some point in the future you can deploy your imagination and demonstrate your values through the vehicle of philanthropy.

By way of example: you are the CEO who wants to root out child labour in your supply chains. This seems the right thing to do; it will reduce your risk, make your supply chains more resilient, and potentially improve your brand equity. But what if your legal, finance, and investor relations team tell you that if it's found in the supply chain then it may lead to lawsuits, or affect the company's credit rating? Do you choose to fit in or defy the status quo? Those around you, in a spirit of conformity, are trying to hold you in the stronghold. You are being told that outside the stronghold is threatening. You need to manage by the rules, shore up the walls, minimise short-term risk, and manage the stronghold.

Conversely, Geoff van Sonsbeeck, of House of Baukjen, decided that he didn't want to manage the stronghold and would rather step outside of its fortifications. In his interview with us he said:

> *I think for us, the epiphany was about four years ago. My wife, who's my co-founder, and I were walking the dog and I had just read that every second a truckload of garments is going to landfill or incineration somewhere in the world. We always wanted to work in the sector and do what's right. We knew it wasn't perfect, but we tried to make changes wherever we could. But when faced with the transparent facts we had to make a decision – are we going to walk away from the sector and just let it be? Or are we going to see whether we can actually change the sector from within and is there a systemic change possible to create a more responsible fashion sector?*

For Van Sonsbeeck, returning to the existing business model and making incremental or 'tinkering' changes here and there was not an option.

> *That was not the scale we were looking for. We were looking to deconstruct the model, the way we knew it, and say if we just look at these building blocks separately, the needs that a consumer has and the choke points in the current linear model, how can we do this completely differently without the consumer having to compromise on quality or price?*

Sometimes when the world looks mad, it's because it is, and the solutions we need to make the change and realign it to values-based logic are in plain sight. The power of strongholds is that they rely on exploitation, misinformation, excessive waste, or preying on the dissatisfaction of individuals. They may prevent us from bringing our values into our work, or from making the decisions that feel right in our spirit. But they are powerful, and despite their shortfalls they bring an illusory sense of safety when all around us feels uncertain.

Breaking Silo Strongholds in Our Value Chains

Within the business structures in our existing systems, we determine what should be in our stronghold and what's outside. Often supply chains are deemed to be outside, a commodity that is in constant flux driven by the desire for cost reduction, where the stronghold is protected by procurement and supply chain management teams, and where Tier 1 suppliers are often known but for the tiers beyond that it becomes increasingly opaque. The need to move beyond the existing relationship with upstream suppliers and consider them as stakeholders is deemed to be a danger zone outside the stronghold.

Now consider the requirement of all companies signing up to credible net-zero targets requiring the measurement and reporting of three scopes of carbon emissions. The three different scope categories of emissions measure a company's own emissions and impact as well as its wider value chain. It forms the basis for greenhouse gas (GHG) mandatory reporting around the world. In summary, Scope 1 emissions covers GHG emissions from a company's activities such as their transport fleet. Scope 2 covers a company's indirect emissions such as purchased electricity. Scope 3 is the most challenging, and often the greatest area of impact and also influence. It covers the GHG emissions produced up and down the value chain, such as from suppliers and from the products when customers use them. This means that the greatest area of impact that needs to be embraced to achieve compliance in the emerging regulatory landscape is outside the typical stronghold.

Furthermore, this can only be done when the entire supply chain, not just Tier 1 suppliers, is included. There is a clear Day 1 driver in these compliance requirements, but there is also value generated by having resilient supply chains in the face of increasingly extreme weather events. The value chain is becoming a more complex ecosystem that is more transparent and where there is a direct link between supply chain resilience, stewardship, and brand equity. Value chains and their component parts will be less easy to compartmentalise within the structures of the future.

Looking downstream towards customers in the value chain is an equally important shift in the models of the future. As we've observed, in the old era, post point-of-sale was also outside the stronghold. Closing the loop on your product cycle and moving from a 'take, make, and throw away' linear model to a circular one is a heavy lift and the introduction of reverse logistics to businesses that have only had to consider forward logistics in the old era is a further systemic shift. Most companies are struggling with this and are observing their market being eroded by the rise of multi-billion dollar reverse logistics firms that know how to operate in circular models. These are new era firms that are currently enjoying the opportunity that exists in an emerging secondary market where refurbished products are sold as new, with discounts, and with warranties.

As we will explore in later chapters the very notion of value chains themselves are strongholds. They set out constructs that are designed to ensure maximum value creation for the owners of the assets (treasures/storehouse) of the stronghold. Value chain thinking encourages stronghold thinking. Stakeholder capitalism opens up the opportunity to rethink value chains as more nuanced, sophisticated, and interconnected systems of value creation and this more open-minded reapproach to value chains will be crucial to the successful journey into Day 2.

Reality and Fantasy in the Midst of the Charge

All this talk about non-conformity, shaking strongholds, and deconstructing the norm will be giving rise to understandable anxiety. As leaders, we need to decide whether we hit the comfort zone 'off ramp' or try to lead through the danger zone. In contemplating this prospect we recommend planting your feet in reality.

We all operate at the frontier of reality and fantasy. Reality causes us to face up to our boundaries such as climate science, planetary boundaries, and sourcing from finite resources. Fantasy, however, promises an extension of these boundaries to postpone the need to address them. Fantasy provides the opportunity to convince ourselves and others that we can remain in the current era, the current stronghold, without stepping into the liminal space of the transition zone. Fake realities will shape fake people and will result in fake leadership. The more we deny the truth that the climate is in peril, the more we allow fantasy to become an ever more potent force in the future destruction of existing systems. This remains a concern with the current denial and cancel culture driven by social media that seems to prevail, alongside the tolerance of fake news, as it supports the confirmation bias needs of the comfort zone.

For example, in the fossil fuel stronghold, the fantasy realm will allow ongoing extraction and burning of fossil fuels beyond the transitional necessity of the targeted 1.5 °C warming scenario. This is driven by climate denial fake news, which will keep us on a catastrophic 3–5 °C global warming trajectory. This could extend the lifetime of this stronghold by another 10, 20, or even 30 years, and allow leaders the option of the off-ramp into their comfort zone. But at some point, there will be recognition that fossil fuels are becoming stranded assets and the stronghold will fall. Facing reality now, however, will take us as leaders into the liminal transition zone. This will feel dangerous, but it will provide the opportunity for transition into the hydrogen economy, carbon capture and the production of bioplastics, low-carbon fertilisers and biofuels, and to participate in the next industrial revolution, possibly even the next stronghold – clean energy.

Decarbonising your operations, moving from linear to circular, becoming inclusive and transparent in the way you conduct your operations and embracing stakeholders who have previously been outside your stronghold is complex. This will not be a linear transition that aligns with the process-driven, highly structured approach of the average strategy consultant or behaviour change expert seeking incremental nudges in improvement. This will be a root and branch reapproach on how you thrive in a non-linear model. This change is not happening in a sequential, or rational way.

We are emphasising this point as we need adaptive leaders who are dynamic enough to seek catalytic sources of activation, operating in a complex system that is non-linear, and where courageous decisions are made at a frontier that is defined by truth.

The examples in this chapter demonstrate the complex web of drivers causing many strongholds to shake, and it is probably the sheer number of them that has broken us out of the old era into this transitional space where we find ourselves compelled to answer– 'How do we find a new path?'

For Richard Mattison, he emphasised that the answer to the 'How' often lies in the information we can already access:

> There's a lot of evidence about supply chains already in play. I think it's beholden on companies to understand what intelligence is available and to use the best of that intelligence to dig deeper, because the process of digging deeper and understanding more about the impact on your stakeholders, and particularly your supply chain as well as your products, leads to a more enlightened approach to running your business. It ensures that you have a strategy that is adaptive to the future and transformative in terms of identifying opportunity.

Scassau *a Muntagna!*

Let us be in no doubt our planetary systems are breaking.

We have expelled this reality from our lives by ousting any manifestation of them from our places of security. We call them externalities as they are external (and hidden) to us when we dwell in our strongholds. But now you have some decisions to make. The evidence is unequivocal – it is now or never. These externalities have the power and force to destroy all our strongholds. If we are choosing to dwell in fantasy to provide the comfort of conformance, then we need to weigh this with the alternative.

Find the reality zone, protect the truth, believe the scientists when they tell you that a lava stream is heading towards your house, and move. Move, both with an appropriate sense of anxiety and urgency but also with a spirit of adventure.

References

AXA Investment Managers (2022) Responsible investing. Retrieved 6 December 2022, from https://www.axa-im.co.uk/responsible-investing

Carrington, D. (21 July 2022) Revealed: Oil sector's 'staggering' $3bn-a-day profits for last 50 years. The Guardian. Retrieved 6 December 2022, from https://www.theguardian.com/environment/2022/jul/21/revealed-oil-sectors-staggering-profits-last-50-years

Chen, Y.H., Landry, E., and Reilly, J.M. (2022) An economy-wide framework for assessing the stranded assets of energy production sector under climate policies. Climate Change Economics. Retrieved 6 December 2022, from https://doi.org/10.1142/S2010007823500033

Courtice, Dame P. (18 August 2022) Interview with the authors.

De Bettignies, H.-C. (September 2017) Developing responsible leaders in China with a global context. The Macau Ricci Institute Journal, p. 78.

De Bettignies, H.-C. (18 August 2022) Interview with the authors.

Douthat, R. (28 February 2015) Opinion: The rise of woke capital. The New York Times. Retrieved 6 December 2022, from https://www.nytimes.com/2018/02/28/opinion/corporate-america-activism.html

Feloni, R. (2021) Despite critics, there's 'no going back' from stakeholder capitalism, says the professor who pioneered the theory in the '80s. Retrieved 6 December 2022, from https://justcapital.com/news/stakeholder-capitalism-pioneer-ed-freeman-on-business-roundtable-purpose-statement-anniversary-woke-capitalism-and-esg-08-2021

Ho, Soyoung. (19 September 2022) More executives could see their compensation tied to ESG goals if SEC finalizes climate disclosure rules. Thomson Reuters. Retrieved 6 December 2022, from https://tax.thomsonreuters.com/news/more-executives-could-see-their-compensation-tied-to-esg-goals-if-sec-finalizes-climate-disclosure-rules/

Make My Money Matter. www.makemymoneymatter.co.uk

Mattison, R. (8 August 2022) Interview with the authors.

MIT Joint Program on the Science and Policy of Global Change. https://globalchange.mit.edu/

Skapinker, M. and Daneshkhu, S. (2016) Can Unilever's Paul Polman change the way we do business? Financial Times. Retrieved 6 December 2022, from https://www.ft.com/content/e6696b4a-8505-11e6-8897-2359a58ac7a5

Stolworthy, J. (23 January 2019) Davos 2019: Bono tells protesters 'capitalism is not immoral – it's amoral'. The Independent. Retrieved 6 December 2022, from https://www.independent.co.uk/news/uk/home-news/davos-2019-bono-u2-capitalism-immoral-africa-aids-wef-poverty-un-a8742071.html

Van Sonsbeeck, G. (8 August 2022) Interview with the authors.

Welsby, D., Price, J., Pye, S., and Ekins, P. (2021) Unextractable fossil fuels in a 1.5 °C world. Nature 597, pp. 230–234. Retrieved 6 December 2022, from https://doi.org/10.1038/s41586-021-03821-8

3

Disruptions
Change as Never Before

In this chapter we will look at the notion of externalities and how these have moved from the periphery of board room discussions to the mainstay of materiality and ESG strategy. We will begin to understand the dynamic and volatile nature as well as the scale of such externalities and the risk they pose to the assets of organisations. We start to explore the nature of stranded assets and the peril for organisations as the drivers of value are gradually or even very rapidly being eroded. We will show how mere compliance with ESG standards is inadequate to protect those assets and even runs the risk of providing false protection against the full power of the disruptions ahead.

Existential Externality

For much of the post-World War II era, growth and stability appeared to be baked into an economic and social model that has provided improved standards of living for most of the world. More than 1 billion people have been lifted out of extreme poverty since 1990. The proportion of undernourished people in the developing regions has fallen by almost half since 1990 (United Nations, n.d.). In the early years, transportation and communication stimulated the belief in the new system of prosperity. Latterly, the technology revolution and the expansion of free trade for an ever-increasing portion of the world's population, has made us believe that we have found an infinite source of wealth creation.

The world has been celebrating the material enhancements of this progress for many decades, and it appears to have been the quiet voices at the periphery that have dared to puzzle why we should expect infinite outputs to come from finite inputs.

An interruption to such a celebrated model that disconnects us from the formula for success, that displaces businesses and markets from their proven source of value creation, is unwanted despite the benefits of cleaner air, less waste, reviving biodiversity on land and in the oceans, improving quality of life, and greater productivity, and so on. All of this is unwanted because of the security we get from conforming to the current strongholds. The world does not like disruption.

And yet here we are. We have been plunged into what seems to be an incomprehensible amount of disruption. What's changed? The Covid-19 pandemic was devastating, a human tragedy. It tested the adaptability of the current model, it brought about new approaches to working lives and operating models, but it was ultimately temporary.

Technological advancement is undoubtedly accelerating, bringing dislocation from the norms of the old era, with increasing transparency of environmental and social impacts in value chains, and power transitioning from brands to consumers. The democratisation of consumerism is changing the power dynamics within value chains and creating a new frontier for competition as the brands built over many decades become more vulnerable to displacement or destruction in previously established marketplaces. Research by Dentsu and Microsoft of more than 24 000 people in 19 countries showed that 59% are willing to ditch brands that do not act enough to stop the climate emergency within a year. And 91% wanted brands to demonstrate they are 'making positive choices about the planet and environment more explicitly in everything they do' (Saga, 2021).

With all these disruptions the malleability of leadership teams has been called upon to overcome radical shifts in the way they do business – from materials sourcing, through to identifying stakeholder and environmental impacts in their value chains, through net-zero pressures in production to communicating more sensitively and authentically to employees, customers, and investors.

While some high-profile corporates such as Blockbuster or Kodak have been on the receiving end of dramatic, sudden, and ultimately destructive forms of technological disruption, technology per se typically doesn't displace businesses from their markets.

So, what's different with the disruptive forces of climate impact and associated sustainability pressures? Why can't the climate be viewed within the frame of the old era? Why are we experiencing multiple catalysts acting as accelerators to the dislocation of the economic, market, societal, environmental realities from our business realities? The mountain has been broken for many decades. Why now?

With climate transition, the frame has moved, both in time and space. Climate change is not a quick fix disruptor through which we can return to the model of the old era. It requires us to decarbonise the world, to find alternative food sources to feed the world's population, to find alternative homes for millions (possibly billions) of inhabitants of this planet, to move from linear to circular models, to stay within planetary boundaries.

But here's the key one: If we don't collectively achieve these outcomes within a narrow time frame, then our planet will be catastrophically broken for those who inhabit it. So, unlike the other disruptions where you can hunker down until after the crisis or make a choice on how and when you embrace technology, with climate, you have no choice. We have a deadline.

The existential nature of climate change and its magnitude of impact is what uniquely defines the climate as a transformational disruptor. We are bound in time, and while we have choices on how we respond, the choice of whether we need to respond has already been taken for us. We must turn the tide within the next 10 years, with a 50% reduction in carbon emissions, and we must transition to a fully 'green' economic model within 30 years.

We have teased out some of the causal agents of this time of transformation in this opening section, but as we focus on the solutions to these challenges, we will see how these drivers of change become the drivers of regeneration, and of replenishment. They will be the drivers that allow us to reconstruct and perform in a new model, and in this frame, we will see the drivers of change as being deeply interconnected. Technological advancement, for example, must accelerate (and become more disruptive), if we are to meet the time pressures of the climate transition.

We'll examine the sharp, short-term, immediate drivers affecting leaders today. We'll look at the reality of what it's like for those who *could* make the change from their places of power, inside individual organisations, but who have many competing pressures. Finally, we'll uncover the opportunity that drivers provide in leading us to the most fertile and exciting time that has ever existed for humanity to leap to a whole new way of life.

Unpacking the Web of Disruption

Long gone are the days when the lone voice of an environmentalist or social reformer would end up pestering the CEO for a Chief Sustainability Officer role – these are now part of the fabric of corporate change. Also long gone are the days of the CEO earmarking a small budget for sustainability reporting and visions, or a risk-reducing CSR strategy with some nice third-party endorsements that keep the

NGOs happy. Over the last decade, with leaders paving the way, sustainability has been woven into the very fabric of corporate life.

This is obviously on a spectrum with laggards at one end and leaders at the other. But there is a constant: every business and every organisation now needs a strategy; at the very least one of harm reduction and at the very most one that's fully regenerative, net positive, measurable, and shareable.

The key indicators to watch out for are how seriously the words are taken and how measurable are the actions delivered; but the drivers are present in a way they *never* have been before and in a way that everyone should be aware of.

Inevitably the upending of the logic of the previous era and the driving force behind the disruption of this liminal space is the climate transition. This will show up throughout this book and also in your day-to-day in terms such as net zero, ESG, circularity, clean energy, green economy – you know the language of 'The Charge'.

But there are other places where we want to shine lights. There are some dark recesses of the current model that need illuminating and as we do so, there is a reaction required from all of us. Incidentally, if you were hoping that at some point in this book we would reveal the silver bullets, then we should probably take this opportunity to disappoint you. The macro driver may be climate but there's a complex web of drivers that sit underneath, which is defining disruption to your regulatory landscape, access to capital and talent, brand equity, customer behaviour, transparency, governance, and risk.

For leaders to be able to react to the size and speed of the challenge, their choices need to shift and types of decisions need to change.

Reacting, responding, and putting strategy in place so change actually happens is far from easy. In a fast-paced world with constant time pulls, we understand the competing pressures. Another day, focused on quarters, margins, and profit, the needs of management of today often strip leaders of the time they need to absorb, process, and think.

It is why we have seen so many goals, pledges, and promises stretch into the 2030s, 40s, and 50s. Why solve today what you can put off until tomorrow? As long as there's a nice headline and some manageable PR flack, it's all good. But things have changed and are changing, and quickly.

Up until a few years ago, the sustainability community spent a lot of time shouting into the wind. As with many aspects of humanity, telling people about change that's coming but hasn't yet arrived, and the need to adapt to it, falls on deaf ears. This has changed forever.

Every day, our media channels are filled with sustainability-related live stories. We are no longer in a world of activists, scientists, and consultants presenting PowerPoints and white papers on what *might* happen in a climate-changed future, we live in it.

This has changed conversations completely. Every conversation we now have is in a reality context. When we talk about carbon reduction, there are fires and floods to relate to. When we talk about regenerative agriculture, there are droughts and polluted waterways to relate to. When we discuss ocean protection there are overfishing stories to relate to. When we discuss human rights there are fashion brand sweatshops to relate to – the list goes on.

It is perhaps the biggest driver of change in terms of the narrative we are all living through. It doesn't always lead to action, as the systems we are trying to change are ingrained and stuck in existing models, but the passion and desire is now much more present and no longer just stemming from the few visionary pioneers who saw it coming.

Casting Some Light

Purpose

Why do you matter? Corporations used to be established with clear public purposes and in some parts of the world they still are. In Chapter 8 we will give some examples. Investment was made to facilitate the delivery of a specific mission. Now we expect businesses to exist to make money and investments to be made to make more money. But profit is the consequence, the product of purpose. Where no genuine purpose exists, the level of trust in corporations is so low that it takes a long time to get stakeholders to commit and believe in the stated mission. But where purpose is soaked in authenticity, it inspires, it gives meaning, it is a driving force for change. It has real power. Why? Because it gives an answer to the why you matter question and an answer to all your stakeholders as to why they matter. A company can have any kind of purpose – an intent to deliver something that underpins its reason for existence – and it doesn't have to be connected to sustainability. However, a purpose is at its most useful when used to connect an organisation's authentic values to delivery via rigorous, meaningful, and measurable outputs. It should be the driver that makes daily decisions easy, encouraging three key questions:

1. Does this decision deliver on my purpose?
2. Does it ladder into my sustainability commitments?
3. Does it stem from my values?

When all three answers are aligned, the decisions leaders make are grounded and clear. Without alignment, decisions become more difficult and messy. It's why purpose is so important as a North Star guidance system, a checking device, and a momentum builder.

For those of us working in sustainability, purpose comes with a capital P. Our version of Purpose is simple: it's the positive impact you want to create from the actions of your business translated into something anyone can understand.

Everyone it employs goes to work to deliver nutrition-based products that improve the health of the people and communities it delivers to. The research and development (R&D) leaders of supply chains and products will likely be transitioning to organic and regenerative resources. You could extend the concept of nutrition to the health of soils, so all the packaging needs to be compostable. As a 'right' it could mean a social nutrition programme that provides its products to people who can't afford them, supported by those who can through a nutrition improvement platform. These are just a few indications of how the Purpose influences the decisions leaders have to make.

The power of a proper Purpose should ripple throughout the business, from top-down at CEO level to bottom-up from the supply chain. Across every aspect of the business's value chain, the Purpose should be living and breathing, thereby ensuring its authenticity, and mobilising energy amongst its stakeholders.

Values

Values and how they drive your business or organisation are a vital part of the sustainability shift. All businesses and organisations should have foundational values: the core beliefs of why they exist, the drivers of their purpose, and the reason that they create impact in the world. We also operate inside the context of the values and norms that society holds and what we culturally, at national and international levels, subscribe to.

Cultural norms shift and change with time and increasingly impact how leaders have to make decisions. Make a poorly informed or ill-advised values-less decision, that doesn't reflect where society is, and there will be some kind of explanation required very quickly, if you haven't already been cancelled. Whilst some of this goes too far, the pendulum swing has taken us to a new place in many ways, with lots of powerful and progressive movements paving the way for leaders to reflect, support, amplify, or expand.

We have seen social movements spring up and take hold in society, from Black Lives Matter to MeToo. These changing values deepen our collective progression towards societal equity and justice, with marginalised groups recovering, discovering, or rediscovering voices anew, that when heard make all our lives richer, deeper, and more inclusive.

This is, of course, a very 'Western' viewpoint in many ways, with fresh challenges around freedoms emerging at the same time. But this is why it is so vital for leaders to know their organisation's values and live by and through them, because without them the seas of opinion can get very rough, with the potential for the boat capsizing.

Customers, stakeholders, and citizens all need to know what is driving you and your business or organisation, how you are showing up, what you stand for, how you will fight, and what you will protect. These are all the types of questions that come with the sustainability shift.

Your values determine how you answer them and should be a key driver of all decisions in and around your business. As leaders, values determine decisions and show up in our behaviours, and we all need to check if our values are in line with the needs of the great sustainability transformation.

Generations

If you are a leader with multi-generational staff you will be all too aware of the need to reflect their needs through your strategic shifts to deliver greater sustainability. The increasing awareness, passion, and influence of each passing generation is remarkable in the authentic change they demand. Millennials began the move, at least the more enlightened ones, and now we are into Generations Z and Alpha. Both were born into a rapidly changing world, with 24/7 global connectivity and are fully digitally native. These are generations that you can't hide from and they demand that the organisation that they work for matches their values.

They typically want to know that their time at work is being spent inside companies that are actively doing their best for sustainability as a minimum. They live in a greatly increased culture of acceptance and diversity and want this reflected. They live with the constant news of climate change and its associated impacts and they want to see authentic responses to this threat.

They are aware of the levels of social inequality that will lead to disruption and they want to be part of the solution and drive diversity and inclusion in society and in their workplaces. Most encouragingly for all of us, the leaders that emerge from these generations will not accept anything but *real* change, and they look up to their predecessors with critical eyes.

As Gina Mastantuono, CFO of ServiceNow told us, 'The new generation of talent is very, very focused on this. They believe unequivocally that climate change is real [and] it's not just a shareholder issue, I hear about it (sustainability) all the time, first hand, from my ~20,000 employees, and customers' (Mastantuono, 2022). This is a very powerful generational wave of change, and it will only get stronger and more determined as their world and future continues to be affected.

Enterprise

Having been in sustainability for 30 years, things have never been more ripe for realising transformative potential. We have identified six factors that are coming together all at the same time:

1. The reality of the change imperative with a looming deadline
2. The will to change
3. The maturing of technology and the power of artificial intelligence (AI)
4. The movement of money
5. The integration of science
6. The demographic shifts and the generational leverage.

The opportunity provided by this web of disruption should be the domain of enterprise. The businesses that are willing to disrupt themselves before they get disrupted will find unparalleled value to harvest. We predict that the energy and appetite of pioneers, entrepreneurs, startups and brave leaders will create a dynamic that will only further amplify the disruption for the leaders waiting to see from the sidelines.

Technology

Invading every part of the disruption web is technology – usually our ally, being at the heart of almost all promising solutions in the transition zone, but sometimes also problematic, such as its energy demand. Wrapped in purpose and believers in enterprise, we see an explosion of incubators, accelerators, and entrepreneurs – pioneers who are focused on solving sustainability challenges, who are hardwired to want to change the world for the better; they simply can't do anything else.

As the founding CEO of Worn Again Technologies, a circular economy textile recycling business, Cyndi Rhoades told us:

> *When we started, we had to explain to potential clients they had a problem and then try and sell them a solution they didn't yet think they needed, that we hadn't yet built; and also try to raise money against that. It's now 15 years later, and after an original investment from H&M Group, the business has developed technology that will enable worn-out pure polyester and poly/cotton blended clothing and textiles to be broken down to virgin equivalent materials for re-entering supply chains to make new textiles, as part of a continual cycle. A small-scale demonstration plant is in development in Switzerland, which will provide a blueprint for scale up and replication, supporting the company's vision of having 40 plants operating at 50,000 metric tonnes per annum by 2040. And this isn't the only technology in the market. There are a handful of fibre-to-fibre recycling solutions at various stages of development which aim to commercialise over the next 5 years,*

thereby playing an integral role in advancing the textiles industry towards full materials circularity.

Rhoades, 2022

What started as a lone pioneer relationship, supported by the first ever start-up investment from H&M, has now turned into an innovation and investment platform, with H&M creating an ecosystem of next generation technologies that can transform the industry. They aren't alone. Industry transformation platforms grow, capital blends from different sources, big business and brands become long-term customers, and consumers demand change; the potential for a whole industry to change to a fully sustainable model is close to reality.

Technology disruption is not confined to small businesses. The R&D centres of companies like PepsiCo and Coca-Cola are still churning out billions of bottles a year. Coca-Cola alone makes 200 000 bottles a minute, which explains their enthusiasm for new technologies to come online, enabling them to pivot away from a reliance on virgin plastics (and therefore oil as a resource). Their supply chain has incorporated the incentives to invent and their investors have the reason to finance. Prior to this there were no real drivers to change – Greenpeace making a noise wasn't enough to give anyone at any point in the value chain a real reason to change anything. Now, everyone has a reason.

Technology is often the vehicle for these break-through innovations that reach beyond the current planetary boundaries – literally in the case of plans to beam solar energy from space. The Space Energy Initiative (SEI), a collaboration of industry and academics, plans to place a constellation of very large satellites in a high Earth orbit.

Once deployed the satellites would harvest almost unlimited solar energy and beam it back down to Earth. According to Martin Soltau, the co-chairman at SEI, 'In theory it could supply all of the world's energy in 2050' (BBC, 2022).

Such technological disruptions even show how human ingenuity can literally and metaphorically transcend finite planetary boundaries.

Artificial Intelligence (AI)

We give AI a separate category to technology, given its potential application and disruptive tendencies. From AI-created art to a system that accurately predicts protein structures changing the entire potential of human biology, from energy-reduction machine-learning efficiencies to mental health support bots, AI is here and integrating at a speed we can't begin to comprehend.

This has huge implications for the decisions that leaders make now and in the coming years, and will affect the way we all create products and services. But how might it help to drive Sustainable Performance? The most obvious is in supporting the incredibly complex data crunching that needs to be conducted to build an accurate understanding of the challenges we face through the implications of models of the future. When assessing possible interventions and inventions to enable the change we need, stakeholders and leaders need to see the future from a place of real data extrapolation. That requires a lot of data, incredibly fast computing, and huge levels of calculation.

If you take a city, for example, the levels of data to truly understand what's happening to it are unfathomably complex.

It sits inside a biosphere that is changing at the speed and scale we have previously mentioned, with all the human activity happening inside it. To understand how the heat, water use, food supply chains, and energy needs might change within that city, we need to model it with huge quantities of data. Before machine learning, creating this type of modelling to any level of deep understanding or accuracy would have been impossible. Now, however, we can break down the data into all its constituent parts and we have access to multiple data sources both from the ground inside the city and off-planet from satellites. We can build a complex model with multiple dependencies and see what happens to the city over time by playing with the models. Spoiler alert: none of it is going to a happy place unless we make significant changes to the way we work and live.

To survive, cities are going to need all kinds of different services and protection and adaptation interventions, as well as cooperation between government, business, citizen organisations, and the people who live there. What AI modelling can do is create known needs from the unknown – these are opportunities for enterprise innovation, highlighting just one AI use for sustainability. Many cities are already using modelling software to explore different greenhouse gas reduction actions, net-zero operational carbon, net-zero embodied carbon and circular economy principles. It is also likely that the management of future cities will be conducted, partly or fully, through AI platforms.

In other areas we can become increasingly ready for climate catastrophe. In Australia, trials are running using an AI system that takes data from forest floors and combines it with satellite monitoring of air temperature and humidity to pinpoint potential ignition points for forest fires. Firefighters can use this information to remove the fuel before the fires start and in doing so massively reduce the size of the fire. These kinds of AI-enabled predictions will hopefully protect lives and property and is indicative of the many opportunities that emerge from climate-related foreseeing.

Robotic efficiency is another area where AI has massive potential. Some leaders at the top of decision-making trees are being asked the most fundamental of all questions: how do we provide energy or how do we feed people? With increasing weather impacts from changes to the climate, agricultural and energy adaptations are just two areas where robotics and AI create amazing opportunities. Whole farm systems could potentially move to robotic regenerative solutions, while underground and overground weather-protected spaces for new nutrition ideas are already flourishing around the world. Whatever happens, the genie is well and truly out of the bottle. AI is ready to be deployed, the only question is what challenges leaders will direct it at.

Impact Accounting

All this talk of disruption, change, and rebellion against the order of the current model is perhaps nothing if we can't measure it. How do we get to the truth of actual performance? That is, performance that measures the creation and destruction of value against environmental and social impacts and not just financial performance. Without this mutinous uprising from the accounting order of today it is hard to imagine how the transparency that investors, employees, and customers demand will be satisfied. And to reinforce the point that these drivers are interconnected, then it is the power of data processing and AI that brings the prospect of impact weighted accounts (IWA) into reality. IWA is the missing piece in the impact economy.

In Chapter 2, we described a reimagining rather than the destruction of capitalism. Now we need the tools to empower this era of capitalism where we can factor into our decision-making the consequences of our business activities not only for financial and physical capital but also for human, social, and natural capital.

In their Harvard Business Review article, 'How to Measure a Company's Real Impact', Cohen and Serafeim (2020) provide the results of work done by the Impact Weighted Accounts Initiative (IWAI) on what happens when you apply new impact rules of the game to the accounts of 1800 companies. Analysing this dataset for 2018 through an impact lens

> brings a new perspective on the true profitability of companies. It becomes apparent that many companies are creating environmental costs that exceed their total profit (EBITDA). Of the 1,694 companies which had positive EBITDA in 2018, 252 firms (15%) would see their profit more than wiped out by the environmental damage they caused, while 543 firms (32%) would see their EBITDA reduced by 25% or more.
>
> Cohen and Serafeim, 2020

But business also generates positive impact through their products and employment, which don't currently drop to the bottom line.

Take Intel's employment impact as an example. In 2018, it created approximately $3.6 billion of positive impact in the U.S. through the wages it paid and the jobs it provided in areas of high unemployment. Intel can increase this impact by improving its level of diversity and offering more equal opportunity for racial minorities and women to advance within the company.

Cohen and Serafeim, 2020

This is an example of what we'll explain further in Chapters 8 and 9 where we describe a Total Value System as a Day 2 shift showing the differences between a traditional value chain and a Total Value System.

With impact weighted accounts, it becomes possible to only tax those companies that pollute to remedy their impacts. Investors will price environmental and social impact into their investment analysis which will, in turn, drive more capital to less impactful businesses, accelerating the rate of change. In fact, the IWAI report already sees such a correlation in several industry sectors.

How close are we to embedding impact weighted accounts in our day-to-day accounting? As we write this, the International Financial Reporting Standards (IFRS) through their Sustainability Standards Board are deliberating on the standards required for companies to disclose their sustainability-related financial information. These proposals are not in response to the NGO or activist community but to requests from the G20. We also await the US Securities and Exchange Commission (SEC) on their decision on the extent to which they will require public companies to make more transparent climate disclosures and provide investors with consistent, comparable, and decision-helpful information.

We all have many things we can spend our time on, but when you have to measure it for reasons of compliance, then it will get the focus.

The Will to Change and Alignment with Capital

We now live in a different sustainability reality where the drivers we have already discussed are a daily presence. This reality enables leaders to have more meaningful, more constructive conversations around sustainability. There is also a real will among boards and stakeholders to make new kinds of choices, opening up even more opportunities for progress towards a sustainable future. Combined with investment pressure and investment opportunity, the landscape for Day 2 innovation is expanding.

The battle for a sustainable future for all is still very real, but the drivers of reality, will, and money can shape spaces we haven't yet seen in terms of innovation.

Let's take a small hypothetical example to illustrate. Imagine you are a product producer that is part of a giant global business. You have a factory in a remote island location, within a community that partly relies on it for work. It is energy intensive and there are currently no renewable options. At a central level, this is a carbon challenge – it is impossible to move beyond limited Scope 1 reductions, and Scope 2 doesn't exist. The solution to reducing the business's carbon and helping to hit its net-zero goals, is for the island to have its own community-owned wind farm.

The head of R&D, with an empowered innovation mandate (will), presents to the CEO using real-time data of the climate impacts (reality) and has worked with the CFO of the business to create a new blended finance option from the market (new money). The business uses its power, talent, and resources to fast-track and organise. It builds the wind farm, co-owns it with the community, powers its factory with renewable energy, reduces its carbon footprint, and has a fantastic brand story. Everyone wins. The resilience of the business's supply chain improves. The factory and its new wind farm improve the resilience of the community and provide a new community micro-revenue stream. The global business reduces its carbon output and gets a great story.

This example could be applied to any kind of intervention in any system. All it takes is for leaders to think differently to change the systems they work within, and thereby lift them to better, more sustainable futures.

The pace of change and the impending deadline for the climate transition may make patience less virtuous. Tweaking BAU won't cut it even if all you want to do is comply. Linear solutions don't work well in a complex web of problems. The compliance stick is out and it is driving us to circular models. Disruption is prevalent at the frontier of technology and sustainability, but it's ripe with opportunity. The world is legislating for a move into the transition zone. It is landing in regulatory pressure and investor and employee activism; consumer democratisation is determining market share, and your licence to operate is being laid bare by the unveiling of the truth within your supply chains. You are in 'The Charge'.

You may believe that your corporate stronghold will withstand the force. Or perhaps you are paralysed by the sheer magnitude and complexity of the required response. Perhaps fear is playing its part as you seek solely to survive.

You will have many questions:

- How do you protect your assets if you step out of your stronghold?
- If value will be lost, how do you extract and create new value in the transition?

- Where do we find opportunity in this transition zone and is it with the start-up entrepreneurs with their VC backers; are they the victors who will gain the spoils?
- How do we disrupt ourselves and take all our stakeholders with us, before we get disrupted?
- What does reinvention look like?
- . . . And what about me and my leadership team – are we up for, and up to this challenge?

The uncomfortable truth is that the engine of our economy needs to be changed while in mid-flight.

There is much to reflect on as we head to the Campfire.

References

BBC, (29 November 2022) How solar farms in space might beam electricity to Earth. Retrieved 9 January 2023, from https://www.bbc.co.uk/news/business-62636746

Cohen, R. and Serafeim, G. (3 September 2020) How to measure a company's real impact. Harvard Business Review. Retrieved 7 December 2022, from https://hbr.org/2020/09/how-to-measure-a-companys-real-impact

Mastantuono, G. (14 September 2022) Interview with the authors.

Rhoades, C. (24 October 2022) Email to the authors.

Saga, E. (7 December 2021) Majority of consumers will 'boycott' brands not acting on climate change. The Media Leader. Retrieved 7 December 2022, from https://the-media-leader.com/majority-of-consumers-will-boycott-brands-not-acting-on-climate-change/

United Nations (n.d.) We can end poverty. Retrieved 11 December 2022, from https://www.un.org/millenniumgoals/poverty.shtml

Part II

The Campfire

We've experienced the 'Charge'. There's no denying its reality, the threat, and the fear of the experience.

What we thought was a space that could be tamed, that was exciting yet secure, is now seen from a different perspective. It's been a seminal moment. Is it a one-off or has the landscape changed for good? We see the foundations of our existence now under threat.

The concept of status-quo is starting to crumble. Is there an alternative to the ever more stringent management of risk in our now disrupted world? An alternative to the rigidity of a compliance response? How might we reapproach the source of the threat in a way that is enriching, where we find new sources of value?

We seek solace around the campfire, to contemplate our new reality, consider the implications for us personally and professionally, and for our organisations and teams, and start to reimagine an alternative future.

4

Reflections
The Insufficiency of Compliance

Strategic Choice

The counterpoints of two opposing positions, situations or approaches as options in organisational strategy is familiar territory. We are used to binary choices where leaders of organisations can choose between two strategic options, usually characterised by varying levels of risk and reward. Often the implication is that there is an easy, short-term and more expedient option and a more nuanced, riskier but more adventurous choice. One is generally positioned as business as usual and the other, in the title of Anita Roddick's book about the pioneering impact enterprise she founded (The Body Shop), as *Business as Unusual*. But does the latter offer a more intuitive or counterintuitive option?

In the Day 1 experience of being charged by an elephant and doing what the guide tells us to do, playing by the current rules, we live to tell the tale – we survive for another charge another day. In the case of this book, these are the rules of sustainability compliance. By Day 2 we consider how to reapproach the sustainability challenge and turn it into an opportunity, not just to survive but to live more fully, to thrive.

In the tradition of binary choice strategy literature, writers have treated us to a cornucopia of metaphor and image. We can swim in 'blue oceans' (Kim and Mauborgne, 2005) of new opportunities or perish in the bloodbath of red ones. We can play 'finite games' (Sinek, 2019) defined by winning in the given frameworks of limitation and rules, or infinite ones, with the promise of continuity and boundless play. We can imagine 'competing for the future' (Hamel and Prahalad, 1996), a longer-term time horizon that we imagine, foresee, define, and create – or we can just plan for how we will compete for success in the near-term calendar or fiscal period.

Poetic metaphor to one side, in each case we are presented with a binary choice: the first option is generally characterised by strategic myopia, a dearth of vision set in the perilous context of a cluttered competitive landscape. The second offers fresh vistas, new frontiers, and abundant opportunity and territory for exploration.

As we reflect further on the nature of the choice facing leaders in our Day 1 and Day 2 metaphor, we are suggesting a choice between a sustainability compliance agenda (to win a licence to operate) and a Sustainable Performance agenda that seeks to build value in a new emerging context for organisations.

Perhaps the most fundamental difference between the classical strategic binaries we've just set out is that those choices operate at an organisational, category, and industry sector level. They concern themselves with questions of our value chain or operating business model. In some of the more powerful and memorable examples given in such literature we learn how it all went wrong. The video/DVD rental company Blockbuster was undone by Netflix due to a value chain fixation on revenues from late return fees, despite already knowing that online streaming was the future for the industry. Or how the clockwork precision of the Swiss watch industry was outpaced by digital alternatives from Japanese entrants, and how Spotify eclipsed the market dominance of Apple's iTunes.

In all these cases the implications for the organisation in question, the category, and indeed the customer experience and investor returns were seismic. But the implications for the wider world were negligible and immaterial. Who even knows or really cares that Blockbuster ever existed? Such strategy is existential only in the narrow sense of the survival of the enterprise in question.

The difference in how an organisation integrates sustainability into its model is critical for the survival of the organisation itself and also in determining the future of our world. The choice is truly existential, and the stakes are higher.

In fact, we can argue that the binary decision-making processes that are playing out are so significant that they are already happening at two levels. What we call the micro-binary – considering the organisation, its value chain, and business model and how it impacts the world. This consideration is happening in parallel to what could be described as the macro-binary, a broader global public debate on what we collectively want business organisations to do, what the levers of change are, and how the outcome of the debate will impact our world and our future.

The relationship between these two planes of action matters. Companies that are setting out to do something different may develop their own theories about what this says about the macro-economic order. The founder of Whole Foods Market, John Mackey, wrote about his motivations and experiences in setting up the business. He tells a powerful founder story of how he and his wife sunk all their wedding gift money into an organic food store in Austin, Texas, much to the alarm of their respective parents. After a year or so in business, the store was hit by storms and floods and all the stock was ruined. 'The flood basically wiped us out. We had no savings, no insurance, and no warehoused inventory. There was no way for us to recover with our own resources.' They were about to go under, when one morning they turned up at the store and found their customers and neighbours had turned up with all the necessary material to clear up the mess and get them back on their feet. Mackey also tells of 'an avalanche of support from other stakeholders as well, all of whom pitched in to save us' (Mackey and Sisodia, 2013, pp. 5–6).

All their so-called stakeholders – meaning the people who mattered to them and to whom they mattered – were with them in this time of crisis because for them the store was meaningful. The stakeholders, who were the real sources of value to the business, wanted them to survive. Mackey decided to frame his experiences at Whole Foods and his foundational beliefs as 'conscious capitalism' (which he explains in his book of the same name), extrapolating his unique approach to managing his own business and its assets to define principles of 'conscious leadership' with broader applications. Those principles aim to show how we can retain the strong driving force of capitalism, the energy of getting started, raising finance, following our dreams, but qualify and even dignify these by higher order motivations of purpose, impact, altruism, and, yes, idealism.

Others who see the advantages of private sector agility and adaptability but miss an inherent focus on impact in their operating model might seek to organise or formalise the collective transition pathway to Day 2. The Founder of the B Corp movement Andrew Kassoy memorably talks about capitalism 2.0, a kind of updating of capitalism software to account for the needs of wider stakeholders beyond shareholders. The B Corp movement has grown to over 6000 companies who are already pivoting from compliance to Sustainable Performance. Part of this pivot involves stating in their articles of incorporation that they exist to serve all stakeholders, not just shareholders, and that they are guided by a spirit of cooperation and even interdependence.

There is another important aspect to consider when contemplating the Day 1 or Day 2 choice, and this is about the fundamental desire, albeit differently expressed in each of us, for our work and our enterprise to be meaningful for ourselves and for the wider world.

The last decades of accelerated and scaled globalisation have brought an increasing realisation that private sector organisations exist in an intricate set of international, social, political, environmental, economic, and public systems and spaces. That world is made up of not only the planetary systems that sustain life but also the intricate web of human interactions that make up our global community. This world, as a composite of planetary and social systems, is a rich treasure chest of opportunity to create enterprises that make a meaningful and measurable contribution to the improvement of quality of life.

Leadership Dilemma: Day 1 and Day 2

As leaders contemplate the dilemma – stay focused on and bound by Day 1 or launch the canoe on and into Day 2 – it may be helpful to acknowledge and identify the types of leaders considering these choices and the kinds of questions which could be running through their minds.

Firstly, they will reflect on the actual lived experience of Day 1 – what they and their organisations must manage and operationalise to make it through Day 1. This is essentially an agenda of increasingly high organisational effort and resources, deployed with ever-increasing demands for compliance just to be allowed to operate, play, and compete – to stay in the game. We have worked with many organisations where even today sustainability is regarded as essentially a compliance topic, with teams and external consultancies tasked with compiling reams of data and information to fulfil the demands of customers, investors, and regulators.

The main motivation behind the stakeholder groups demanding this level of data-driven compliance is to avoid risk – risk in their own Scope 3 (supplier) value chains, risks to their invested capital and risk to their reputation. In the case of regulators, it may be an increasing impatience for private sector organisations to take action on long-standing problematic practices in their supply chains such as deforestation, unwise linear use of finite resources or issues related to human rights and fair labour practices. Whatever the motivation, we face an avalanche of new laws and standards that require ever more organisational headspace.

Investors and Jargon-Busting

ESG has become part of the vernacular for companies seeking a score to reflect their performance against these criteria.

It is a framework defined by a set of criteria that allows an accessible understanding of the performance of a business (sometimes cities or states). Those criteria measure a company's ability to mitigate against environmental degradation; the social responsibility in its relationships with suppliers, employees, customers, and the communities that are impacted by their activities; and the way a company governs itself in areas such as leadership, executive pay, internal controls, etc. The opportunity that ESG metrics provide to convert performance into a score has made it the reference increasingly used to inform the investment choices of large institutional investors, such as public pension funds.

Impact, on the other hand, reflects an investment class where capital is deployed into businesses whose mission involves generating positive, measurable social and environmental impact alongside a financial return.

Beth Houghton, Head of the Impact Fund for Palatine Private Equity, puts it like this:

> *ESG is how a company delivers its goods and services. So every company in the world can improve its ESG score because it's about how you look after your employees, how you engage with the environment, relations with suppliers, and your governance protocols etc. That's applicable to every business.*
>
> *We invest in companies that create social or environmental change as the core activities of that business. So it's about what the business does, and where we can drive value through that mission.*
>
> Houghton, 2022

There are many other terms such as *corporate social responsibility*, which reflects the commitments a business makes to give back to the environment and society through a model that often supports a culture of philanthropy. *Responsible investment*, on the other, hand reflects the strategy embraced by financial institutions to incorporate ESG into investment decisions.

Layered on top of this increasingly congested set of terminologies is a growing interest in *impact-weighted* accounting, as we outlined in Chapter 3, or *environmental P&L*, and other models that effectively price in the environmental and social externalities of a corporation's assets, such as by placing a value on natural ecosystems or penalising an organisation for polluting emissions. In this way, we are taking the positive and negative impacts of a corporation's activities and pricing these into the P&L and balance sheet, thereby accelerating the reduction of negative impact and increasing the respect for the value that nature provides.

Of all these terms and methodologies, it is ESG that has gained traction in the common parlance of the business community. It is also ESG that is often misused or overused as a proxy for sustainability strategy.

Given the world's focus on ESG, it has become the combat zone for the colliding philosophies to battle things out. We remain watchful as to who will be victorious in the design wars for the next era. Writing about ESG in an article for *The Economist*, Henry Tricks argues that:

> *the concept's popularity has been partly fuelled by real-world concerns, especially climate change. Yet it has had a negligible impact on carbon emissions, especially by the biggest polluters. Its attempt to address social issues such as workplace diversity is hard to measure. As for governance, the ESG industry does a lousy job of holding itself to account, let alone the companies it is supposed to be stewarding. It makes outsize claims to investors, and it puts unmanageable demands on companies.*
>
> Tricks, 2022

This stinging critique of a concept that is increasingly dominating the sustainability approach and agenda for business reflects growing concerns about the process of monitoring and reporting for ESG and its limitations for impact. It is easy to see why. For example, no single, global standard for ESG reporting exists; there are a lot of regional or industry-specific standards to choose from and organisations will use multiple frameworks. The regulations and standards for the industry are constantly evolving and the reality is that sustainability is hard to quantify. ESG data is often divided in silos across a business, making it difficult to grasp a global, integrated picture of ESG gains and impacts.

However, Tricks goes on to acknowledge that:

for all its pitfalls, it may be better to overhaul than to bin ESG. At its core, it is a quest for something increasingly crucial in the battle to improve capitalism and to mitigate climate change: making firms and their owners accountable for their negative externalities, or the impact of production or consumption of their products on third parties, such as the atmosphere. By forcing businesses to recognise the unintended consequences of many of their activities, the theory is that they should then have a greater incentive to fix them.

In Chapter 2 we introduced Richard Mattison, who leads S&P Global's sustainable intelligence platform, Sustainable 1. Mattison recognises the importance of ESG and its growing dominance across the corporate world, but also the potential for a backlash.

ESG, is ripe for critique right now . . . The fact it's happening shows [its] importance as a framing tool, but for it to become more useful we need to be clear what's in and what's out. As reporting of ESG performance becomes standardised and auditable then we can enable a more responsible allocation of capital.' He sees pushing for 'greater transparency and objectivity' in the area of ESG as essential and states that 'efficient markets are hungry for data on risk and opportunity, and this is what ESG data is striving to provide.

Mattison, 2022

The interest in ESG is partly fuelled by some high-profile incidents where poor ESG performance resulted in financially material impacts. Examples include the Deepwater Horizon incident for BP, BooHoo's modern slavery scandal, Cambridge Analytica harvesting data without users' consent and the impact on Facebook, and Volkswagen and their misrepresentation of emissions test results. Many billions have been wiped off market valuations, or have been imposed in penalties, fines, or other liabilities.

We accept that ESG is the 'buzz' term that captures the prevailing approach businesses are taking towards sustainability right now. We need the investor community on board as a partner in Day 2, and if ESG acts as a helpful compass for this community then we should welcome it and celebrate its influence. There is no doubt that this data-framed approach will satisfy the appetite of those who seek to tick boxes to achieve compliance and in so doing contribute to the resolution of global problems. ESG, we believe, is here to stay and there will likely be an increased standardisation of this driven by industry needs and regulation.

Regulatory Compliance

In the past two years alone, we have seen the arrival of the EU Corporate Sustainability Reporting Directive (CSRD) proposal to replace the Non-Financial Reporting Directive (NFRD). This will require nearly 50000 companies to report ESG data according to new EU sustainability reporting standards. The IFRS Foundation, a non-profit accounting organisation, announced the creation of the new International Sustainability Standards Board (ISSB) to deliver a global baseline of sustainability-related disclosure standards and, more recently, the US Securities and Exchange Commission (SEC) proposed new rules for climate-related disclosure that would require US companies to report on their greenhouse gas emissions and detail how climate change-related risks will affect their business.

This is the Charge of Day 1 in all its foot-stamping, trumpeting intensity, and for many leaders and businesses, Day 1 quickly becomes Groundhog Day. It turns into a never-ending cycle of data collation and disclosure to fulfil the demands of an equally ceaseless litany of reporting standards and acronyms, including GRI, CDP, SBT, TCFD, SEC, CSRD, FHSA, CPSIA, SFDR Sustainable Finance Taxonomy, EU Taxonomy, EU CSDD FLAG SBTs, among others.

One might expect seasoned, smart, and wise leaders to question the value of much of this compliance agenda and what is driving it. The C-suite should be debating the degree to which the compliance requirements have become the reporting tail wagging the corporate strategy dog.

The cost and requirement of compliance will be high. There is already a regulatory onslaught. The European Union is driving its Green Deal and the EU Taxonomy, the UK has the Climate Change Act, the new US agenda on the SEC/TCFD-aligned Rule and the Responsible Business Initiative in Switzerland. As if the force of this regulatory wave were not enough industries are even lobbying regulators for tougher oversight and legislation. The motivation is to ensure that subsets of companies which are more diligent in compliance can play on a level playing field with regulatory laggards. Or, in some cases, expecting that the pressure to comply will mean some players will ultimately fail to keep up with the demands of the compliance agenda.

In this debate there will need to be a distinction between the range of different types of Day 1 compliance – mandatory regulation and voluntary standards, and the

GRI: Global Reporting Initiative
CDP: Carbon Disclosure Project
SBT: Science Based Targets
TCFD: Task Force on Climate-related Financial Disclosures
SEC: Securities Exchange Commission (US)
CSRD: Corporate Sustainability Reporting Directive (EU)
FHSA: Federal Hazardous Substances Act
CPSIA: The Consumer Product Safety Improvement Act
SFDR: The Sustainable Finance Disclosure Regulation
EU CSDD EU Directive on Corporate Sustainability Due Diligence
FLAG SBTs: Forest, Land and Agriculture Science Based Targets

degree to which non-compliance with respective standards risks access to capital or markets. But there should and must be more attention paid to the meaning and the value in this compliance.

In a recent project for a publicly listed MNC in Europe we discovered that, at C-suite and board level, the company's sustainability function was essentially filling out forms for customers and investors and other reporting frameworks and that the organisation had hardly any additional resource capacity to create a sustainability strategy driven by the business and linked to its corporate strategy. It was striking that when we interviewed investors and key account partners (the company's stakeholders), they were calling for more than compliance. Customers were open to sustainability-driven innovations that would create new sources of differentiation and many investors were looking not just for ESG compliance but impact-driven value creation from their investment. Leaders should be considering how the intensity of Day 1 compliance activity may be holding them back from creating value and embracing sustainability opportunities and strategies.

In this market that is hungry for transparency, customer empowerment, and, of course, the potential for superior returns we shouldn't be surprised that investors have got excited about seeing ESG and Impact as levers or proxies for superior performance. MSCI discovered that high-ESG-rated companies outperformed low-rated companies during a 7.5-year study period in all regions, which was mainly due to superior earnings growth and, to a smaller extent, higher reinvestment returns (MSCI, 2021).

All of these measures are designed to shine a light on performance, which should help inform investment decisions. But the plethora of new terminology is in danger of creating noise that will be both confusing and inaccessible to many. In this book we aspire to co-create a class of business for the new era, not scored for how much they minimise damage, or how green or socially responsible their mission is, but for their Sustainable Performance.

Businesses which embrace Sustainable Performance recognise that an enterprise that operates in partnership with nature, that flourishes within planetary boundaries, that unleashes human capital across the entire value chain, that instils resilience into the communities it serves, and that returns value to all stakeholders, will outperform. For businesses to survive long term, we would argue that this is the only way.

Changing Customer Habits

The NYU Stern Center for Sustainable Business reported in 2021 that sustainability-marketed products delivered approximately one third of all CPG (consumer packaged goods) growth, despite only representing a 17% market share. Products marketed as sustainable also grew 2.7 times faster than products not marketed as sustainable, and achieved a six-year compound average growth rate (CAGR) of 7.3% versus 2.8% for their conventional counterparts. In addition, the report noted carbon-labelled products now account for US$3.4 billion in sales, up from US$1.7 billion in 2020 (Kronthal-Sacco and Whelan, 2022).

This indicates that, whether businesses are on board with the sustainability agenda or not, consumers are increasingly voting with their wallets. It's clear there is a growing market for sustainable products, and a growing willingness among consumers to pay for them.

Things are getting real. The Charge of climate change is gaining unsettling momentum. This, coupled with the Charge of social inequity, is leading to the steady unravelling of social cohesion with subsequent pressure on businesses and leaders to act. This unsettling connects with the shaking of strongholds as we have described. Patagonia's founder Yvon Chouinard's announcement in 2022 to make Earth the company's only shareholder hints at the world to come and the pressure soon to be piled onto businesses and brands to authentically deliver on their promises. We shall say more on his example in Chapter 9.

The way in which we live, work, and run business is not sustainable, and the vast amassing of wealth by a few at the expense of many is ultimately not acceptable. But what does the customer think about this? Are consumer habits really changing in line with the growing pressure of sustainability on business?

We could argue that customers changing their buying habits are outliers, and that the consumerism stronghold is holding fast – people still buy fast fashion and appear to make purchasing choices based on price rather than sustainability. On the face of it, people don't appear to be changing habits and attitudes at any great speed. But perhaps it is the forerunner, or maybe a bellwether for what is to come. Perhaps the consumer is savvier and far more aware than we might think.

According to a recent Deloitte survey, consumer habits are changing in line with a growing awareness of environmental and sustainability issues. Its 2022 study into sustainability and consumer behaviour revealed that 59% of consumers are now buying only what they need, up from 39% in 2021. In addition, 30% of respondents said they had opted for low-emission or shared modes of transport (up from 19%) and 39% said they had reduced meat consumption (up from 30%). Conversely, however, 52% of consumers say it is too expensive to adopt a more sustainable lifestyle and 57% say they would do more if it was more affordable (Archer et al., 2022). These statistics are enlightening. Cost remains a barrier to consumers making decisive choices towards sustainability, but the needle is moving – people are intentionally buying less and have a clear desire and willingness to adopt lifestyles that are more in line with supporting a healthy planet.

Unilever's 'Making Purpose Pay' report further reinforces this data. Interviewing 20 000 people across Brazil, the UK, the United States, India, and Turkey, they discovered that while cost remains a barrier for some, 'over half of all consumers already buy or want to buy sustainably: One in three (33%) already purchases products with sustainability in mind and a further 21% do not currently but would like to' (Unilever, n.d.).

What does this all mean for brands and businesses? The pressure of the Charge is clearly building, and consumers are now seeking out brands and businesses that are taking decisive action in response to sustainability challenges. This means that brands are increasingly required to make decisions and announcements in line with this emerging consumer interest. Purpose statements are being created and honed, sustainability ambassadors employed, and public commitments to targets announced. But this poses a problem that is exacerbated by the Day 1 approach of adherence to compliance, tick the boxes, and create incremental change. Not all businesses claiming green credentials are taking authentic and transformational actions for change.

Accusations of greenwashing are decried by critics as often as sustainability claims are made by businesses, and for the consumer this creates confusion, frustration, and leaves them uncertain as to which claims to believe. In February 2022, financial services firm Morningstar exposed the potential depth of this issue when, following forensic analysis, it stripped the ESG tag from funds managing $1 trillion (Schwartzkopf and Kishan, 2022).

There is no doubt that engaging meaningfully with sustainability and pivoting for a new era is not going to be easy. So many businesses remain stuck in the paralysis caused by the Day 1 Charge, employing sticking-plaster solutions and creating a sustainability mirage in the hope that the customer won't notice. Of course, many customers do notice. Unilever's (n.d.) research reported that 'consumers were quick to pick out the brands they think are simply jumping on the "doing good" bandwagon. So how brands behave is as important as what action they take, and their consistency of message across key touchpoints. Anything less than a joined-up approach risks being seen as inauthentic.'

Yes, there is a lot of confusion for consumers but increasingly the brands that are stepping into the transformational, 'art of the possible' arena are becoming easier to identify and, as the story highlights, the bar is being set increasingly higher. Consumers too are digging deeper into the statistics and claims, looking for authenticity, transparency, and truth. The Deloitte survey emphasises that brands need to work on helping their customers to trust their sustainability commitments, and consumers are increasingly echoing what brands are hearing more and more often – the proof is in the pudding (Archer et al., 2022).

Authenticity and transparency will eventually win through. The data shows that for the brands that create and deliver great quality products that can compete on cost and deliver sustainability messaging and action that doesn't crumble under scrutiny, the consumer is ready and willing to buy. But, of course, this requires imagination, perspective shift, innovation, and new models of working, which only a Day 2 'Reapproach' can bring. For those businesses stuck in the face of the Charge, choosing not to – or feeling unable to – engage in meaningful change, may make reapproaching the challenge feel impossible.

Talent

Workers are winning the war for talent. There's a debate over whether you will find the best salespeople in your organisation in your sales team or in your recruitment team. Talent is selective, it's discerning, and it's hungry to find workplaces where the corporate values align with personal mission. Talent wants to make a positive difference in this troubling world. It can be a mystery to many of the post-industrial leaders that still hold much of the corporate power in the world, that the hierarchy is less able to wield power and that instead of the favour that companies were offering to employees in the past, they now have to roll out the red carpet to get them through the door.

There is plenty of research showing the importance of purpose and sustainability in attracting talent. It is often one of the main reasons for organisations embracing a B Corp status. When Innocent Drinks became a B Corp in

2018, CEO Douglas Lamont declared on Twitter it was the 'proudest day of my career' describing it as a 'pretty effective way to grow your business, attract great talent, drive up employee engagement, and deepen brand love'.

Gina Mastantuono, CFO of ServiceNow, reinforces the connection. 'We're in the middle of a perfect storm – the next generation of talent demand to work for purpose-driven companies and regulations aren't slowing down. The war for talent is real and attrition is costly' (Mastantuono, 2022).

This is the point where we could throw countless data points at you to reinforce the argument that talent attraction is one of the greatest benefits of having an authentic purpose-driven business. It's all true, and you and your teams may be looking for the optimal ESG story to supplement your market sales story as you seek the best talent. This is good, but it's been written about before.

We want to approach this a little differently. Instead of talent attraction and retention being a primary reason for embracing ESG and impact, we want to instead present it as a resultant output of Sustainable Performance. This is a subtle shift, but it focuses us back on the North Star of Sustainable Performance and to achieve this we will need talent who have aligned values and purpose motivations with the organisation, rather than us trying to bend the mission of our business to appeal to the talent market.

To unpack this, let's relook at some of the characteristics of the transition from Day 1 to Day 2. These are the assets and behaviours that you need to move from the paralysis of the Charge to the Reapproach in the canoe.

1. *Suppliers*: Your suppliers are almost certainly global, and culturally and ethnically diverse. With increased transparency, you may uncover abusive or unfair labour practice and human rights violations. There may be ethical

dilemmas that require the north star guidance system and wisdom.

2. *Customers*: Your customers are operating outside your control zone. Technology has given your customers access to competitors and the platforms to declare their opinion on your service, product or brand with little or no control from yourselves. Democratisation of consumerism has arrived.

3. *Loyalty over leverage*: Navigating successfully in this zone will require a different relationship where you need to build trust, respect, and loyalty with your stakeholders rather than seeking to exploit your enhanced leverage to maximise profit with little regard for social or environmental impact.

4. *Connecting with stakeholders*: Your communications department needs to know what authenticity looks and feels like and how ESG claims will stand up to scrutiny.

5. *Manufacturing*: Your product development team needs to design products with recovery and reuse in mind. The assembly team will need to source much of their raw material from the components of existing products that need to be retrieved from customers.

6. *Finance*: Capital will be deployed by demonstrating that you are making the required progress in the transition and meeting milestone ESG targets. Investments will have to be made in reverse logistics. Sales teams will have to be incentivised differently and enterprise management systems and inventories will have to be reconstructed for the new business models.

7. *Technology*: To move from a linear, forward-focused business model to a Total Value System is complex. We need to embrace those tools that will help us operate effectively in complexity.

This is the transitional landscape of behaviour relating to asset and operational management to support Sustainable Performance. So now to the question: As a leader, will you be able to deliver the above with the same leadership team and people assets that were effective for you in the old era?

The answer is almost certainly no. You will need talent to effect the transition to Sustainable Performance. This talent needs to be drawn to the mission of the business and be motivated by the challenge of the pivot that you are seeking to make. The narrative of aligning your ESG values to those of the talent you want to employ is turned around by you setting the standards. In a journey to Sustainable Performance, talent will want to align to the Purpose of your business.

Those companies with the greatest challenge in this pivot, the most adventurous and treacherous journeys in the canoes, are fossil fuel companies, the steel and cement works, and energy intensive industries. If these companies and others – including consumer-facing, resource hungry, externally damaging companies – do not succeed in the Reapproach then the hope and excitement that we might have for the next era will be greatly diminished. We need great talent to be flowing into these sectors, but if the story of these industries directed towards purpose-driven talent is greenwashed, then discerning talent will see through it. A meaningful mission towards realising Sustainable Performance provides the authenticity that should attract the most aspirant of talent in this purpose-driven marketplace.

We are considering here the appeal a business has to attract the right talent. But a word on culture. Clients and customers will have little patience for siloed structures, tribal cultures, and private fiefdoms. These old era forms of command and control get in the way of good business, getting sustainability done, and getting it done seamlessly across global value chains. Bureaucracy and over-reliance on top-down control will be seen to stifle performance. Instead, businesses will want to unleash potential through personal and company-wide empowerment, and an entrepreneurial can-do mentality. They will prioritise pace over perfection. Cultures will be founded on trust and respect, as opposed to cultures that see people as a risk that needs to be managed.

New cultural constructs need new types of leaders, which we shall discuss in detail later. For now, a quick observation: Whether it was Covid-19 levelling the playing field, the climate transition, or other liminal disruptors, we see the shifting relationship with talent is giving more space to our next-generation leaders. This will be a critical success factor at the new frontier. The next-gen leaders will succeed at the digital and social frontiers, which is where the battle for the Decisive Decade will be won or lost. We are the first generation to have the climate impact facts at our fingertips and the last generation to be able to do anything about it. Having the right people around us is a critical responsibility for leaders.

This time of accelerated change is not transitory or irrevocable. This is a time in which the fundamentals are being reset. The digital and social-era models are inherently faster, more agile, and flexible than the traditional hierarchical models that preceded them. Liminal is where new systems will be revealed. Liminal will require us to lead through influence rather than hierarchical positional power.

Perhaps we can shift the power dynamics from hierarchies of status power, to communities of passion, purpose, and collaboration where we share a common interest and mission.

Reapproaching Your Supply Chains

'Finding child labour is a good thing', said the hard-hitting headline on the Tony's Chocolonely ads in 2022, 'because then we can help sort it', said the subheading. The new era is yet to be created but we can see design elements already emerging. One of these is radical transparency. Tony's Chocolonely are a company born as native canoers in Day 2. They call out the brokenness of the value chain, the 1.56 million children working illegally on cocoa farms in Ghana and Côte d'Ivoire. They are proactive in calling out the tragedy of the exploitation models of the old era value chain where cost has been prioritised over human rights. But here's what the company says next: 'This is what we exist to change.' Why does this business exist? Why do they matter? What's their purpose? Yes, they make great chocolate but their ad campaign makes little mention of this. Instead, Tony's is using business as a vehicle to help eradicate child labour.

Now let's ask the question, 'How does the company make it work?' Tony's publishes a programme called the 5 Sourcing Principles. Those farmer cooperatives that use the 5 Sourcing Principles have an average child labour rate of 3.9%, compared to the industry average of 46.5%.

Larger confectionery brands are managing complex supply chains with high volumes and significant numbers of farmers. Many are doing impactful work behind the scenes to reduce the risk of child labour and introduce sustainable farming practices to improve livelihoods. Tony's have taken a more overt stance. As a challenger Day 2 brand they have sought to differentiate their brand proposition by linking it to the most material issues in sourcing cocoa. Tony's claim is that they are solving the structural child labour issues in the supply chain and communicating openly on how they are doing it. Tony's approach is to bring transparency into the value chain and integrate it within the brand proposition as a distinctive feature of the consumer buying decision.

Tony's are stretching their corporate arms around the value chain to bring greater integration and governance across all their activities. They know what's happening so they can leverage their corporate power to do good. What do they get in return? Greater supplier loyalty, yes; but now they are clear on their supply chain-led reason for being, they can travel this story into their brand presence and appeal to customers. Tony's sales growth relies on building brand advocacy and word of mouth. They aim to recruit 'chocofans', or people who love and enjoy chocolate, and to turn them into 'serious friends' of the brand who might give Tony's chocolate to someone as a birthday present, buy a personalised bar, or share the story through media channels. They have moved upstream and downstream in their relationship with their stakeholders to achieve Total Value System integration.

How has Tony's Day 2 transparent approach to child labour affected the company's financial performance? In 2021 Tony's announced that they had broken through the €100 million revenue from barely reaching €1 million 10 years previously. They have a 21% market share in the Netherlands and are now enjoying rapid growth in the UK and other European countries.

Some of you will be reading this and considering those conversations that you've had with your board about lifting the lid on your value chains, taking your customers with you and finding more value in your brand by extending your relationships and sphere of influence up and down the value chain. Perhaps you've had your legal counsel telling you that you will be exposing the company to the potential for lawsuits if something is uncovered, or your investor relations team expressing concerns over the impact on your ESG ratings, or your marketing team raising the alarm on sales or brand value if you go looking in the way Tony's has. Perhaps you are being told that it would be a breach of your fiduciary duty to do something that might jeopardise the interests of shareholders.

They are holding you in your stronghold, and here's the punchline: Transparency will expose you. So, as with everything in life, you do have a choice. Manage the revelation, be transparent with prudence, or lose control in a future where transparency will expose what's happening upstream and consumer democratisation will expose you downstream.

Is There Really a Choice?

It is increasingly clear that circumstances have made the choice for us. The reality is that all organisations will need to comply, but compliance alone will no longer guarantee survival.

There is already evidence that gradually the world is pivoting to Day 2. The strongholds are shaking. The world and planet need more change and innovation than Day 1 can give. As the sustainability industry grapples with a standardisation of ESG definition, indicators, reporting frameworks and methodologies, we worry that this may vindicate those who want to stay in Day 1 to remain in Day 1, with a collective self-delusion that this will be enough to ensure organisational survival. We are setting this against the potential for a Reapproach towards a living world that supports human flourishing.

So, even if we think we can avoid it, Day 2 is the only option, and the question is: how we can get there quickly and at scale and in a way that really does deliver impact and performance? In Part 3 we identify a range of different types of engagement with the reapproach experience of Day 2.

References

Archer, T., Cromwell, E., and Fenech, C. (2022) How consumers are embracing sustainability. Deloitte. Retrieved 7 December 2022, from https://www2.deloitte.com/uk/en/pages/consumer-business/articles/sustainable-consumer.html

BooHoo. (2022) Our vision and strategy. Retrieved 7 December 2022, from https://www.boohooplc.com/group/our-vision-and-strategy#:~:text=We%20will%20continue%20to%20deliver,an%20open%20and%20transparent%20way.

Hamel, G. and Prahalad, C.K. (1996) Competing for the Future. Harvard Business Press, Boston.

Houghton, B. (18 August 2022) Interview with the authors.

Kim, W.C. and Mauborgne, R. (2005) Blue Ocean Strategy: How to Create Uncontested Market Space and Make the Competition Irrelevant. Harvard Business Press, Boston.

Kronthal-Sacco, R. and Whelan, T. (April 2022) Sustainable Market Share Index 2021. NYU Stern Center for Sustainable Business. Retrieved 7 December 2022, from https://www.stern.nyu.edu/sites/default/files/assets/documents/FINAL%202021%20CSB%20Practice%20Forum%20website_0.pdf

Mackey, J. and Sisodia, R. (2013) Conscious Capitalism: Liberating the Heroic Spirit of Business. Harvard Business Review Press.

Mastantuono, G. (14 September 2022) Interview with the authors.

Mattison, R. (8 August 2022) Interview with the authors.

MSCI. (2021) The Drivers of ESG Returns: A Fundamental Return Decomposition Approach. Retrieved 7 December 2022, from https://www.msci.com/www/research-report/the-drivers-of-esg-returns/02343319303

Roddick, A. (2000) Business as Unusual. Thorsons.

Sinek, S. (2019) The Infinite Game: How Great Businesses Achieve Long-Lasting Success. Portfolio, New York.

Schwartzkopf, F. and Kishan, S. (10 February 2022) Funds managing over $1 trillion lose ESG tag at Morningstar. Bloomberg Law. Retrieved 7 December 2022, from https://news.bloomberglaw.com/capital-markets/funds-managing-over-1-trillion-lose-esg-tag-at-morningstar-1

Tricks, H. (21 July 2022) A broken system needs urgent repairs. The Economist. Retrieved 7 December 2022, from https://www.economist.com/special-report/2022/07/21/a-broken-system-needs-urgent-repairs

Unilever. (n.d.) Making Purpose Pay: Inspiring Sustainable Living. Retrieved 7 December 2022, from https://assets.unilever.com/files/92ui5egz/production/f5275ae450e58c008842c928cc2607ea1df2e884.pdf/making-purpose-pay-inspiring-sustainable-living.pdf

5

Pivot
Mindsets, Models, and Honouring Planetary Boundaries

We are moving from a Day 1 situation, defined by compliance and incremental change, to a Day 2 action where new mindsets, technologies, and economic models will intervene in support of the transformation from one era to the next. We know that business as usual can no longer be our modus operandi but what needs to move, change, or take on a different shape for us to succeed?

In this chapter we'll take a closer look at how we extract ourselves from what we have come to see as 'normal'. From the way we have commoditised the natural world, the business models, and asymmetrical relationships that define how we work, the way we produce, consume, and throw away what we make. When viewed through a different lens it will become apparent that these norms no longer fit the world we live in and that we need to reset the foundations for the emerging new era.

Our message in this book is that success in Day 1 and a transition to Day 2 are not optional. Sustainable Performance will only happen if we can make the shift back to operating within planetary boundaries. For us to do this we need to coalesce our supply chain around our global net-zero targets. We need less transactional and more longer-term relationships with our clients. We need to get our product back at the end of its life, retrieve it, and recycle/reuse its materials. A shift like this will disrupt the supply chains as we start to source our raw material from customers. Transparency across the system will mean that brands become much more vulnerable in parts of the value chain where they currently have little visibility. Businesses will need to be highly adaptable to keep up with customers who typically adapt their buying choices faster than corporations can respond.

We can understand the imperative for change and activating the Sustainable Performance of Day 2 by examining how current business models ignore the contradiction between the need for an infinite supply of raw materials, a limited carrying capacity for waste, and the reality that we operate within finite planetary boundaries.

Operating Within Planetary Boundaries?

It is now undeniable that our prevailing systems (strongholds) have been designed to assert control over the natural world. These strongholds wilfully dismiss destructive externalities connected with the ways in which we live and work and have reaped the far-reaching negative impacts we have described.

We do believe that it is possible to find operational and economic models that are infinite, where the process of Sustainable Performance can allow humanity to thrive indefinitely within the constraints of our planetary boundaries. We have shared an outline of an established process we use to help organisations in the pivot towards Sustainable Performance, which we call The Activator Journey, in Appendix 1. But to start, we need to step away from the deceptively stable and finite zones of the previous era into the perceived danger zone of the transition, the liminal space we are now in.

The planetary boundaries concept came from former Director of the Stockholm Resilience Centre, Johan Rockström, who led a group of 28 internationally renowned scientists to identify the nine processes that regulate the stability and resilience of the Earth system (Rockström, 2015).

They introduced quantitative planetary boundaries within which humanity can continue to develop and thrive for generations to come, and where the crossing of these boundaries increases the risk of generating large-scale abrupt or irreversible environmental changes.

These planetary boundaries are defined by nine key processes: climate change, biodiversity loss, ocean acidification, ozone depletion, atmospheric aerosol pollution, freshwater use, biogeochemical flows of nitrogen and phosphorus, land-system change, and release of novel chemicals.

Their identification represents an opportunity to look at how disruptive the externalities are likely to be on our operational and economic models. The processes where we are currently operating outside the safe operating space are: climate change, biodiversity, land-system change, and biogeochemical flows (nitrogen and phosphorus imbalance). By understanding where the planetary stress is we can see where we need to prioritise remedial activities. For example, the processes above where we are outside the safe operating space can lead us to the reimagining of our energy and food systems.

Economist, Kate Raworth, used the principle of planetary boundaries to inform her *Doughnut Economics* model (Raworth, 2017). This model seeks to shine a light on the need for alternative economic models to lay a foundation for the next era. Alongside the planetary boundaries described above, she maps the social foundations necessary to ensure the needs of all are met 'within the means of the planet'. The picture she paints is stark but clarifying.

In 2008 the world observed the slow-motion accident of a world-changing financial crash. Today we are doing the same in the way we disregard nature and its limits; we are wilfully closing our eyes as disaster slowly, and increasingly rapidly, unfolds. By believing that we can achieve infinite economic growth that relies on a limitless supply of raw materials, and air, land and oceans that can accommodate an endless supply of waste produce, we are ignoring the constraints of our finite boundaries.

Raworth's Doughnut Economics model is presented by two circles forming an inner and outer edge. The former represents 12 dimensions of the social foundation (drawn from the Sustainable Development Goals), and the latter, the ecological ceiling. Between these two is deemed to be a 'safe and just home for humanity'. Doughnut Economics confirms the reality that economies, societies, and the rest of the living world, are complex, interdependent systems that should be understood through the lens of systems thinking. It displays the need to transition from degenerative to regenerative and from economic division to economic distribution. In the spirit of stakeholder capitalism described in Chapter 1, Doughnut Economics calls on the human behaviours of cooperation and nurturing that can coexist alongside self-interested competition.

However, as with all new economic models, they have their critics. Some of the ingredients of the Doughnut model are deemed to have been part of failed economic models of the old era. Greater command and control by the state and assets being brought back into common ownership are features of this model that are likely to polarise audiences.

We are not looking to steer the readership of this book in one direction or another in terms of economics preferences, but we do ask for consideration of the context of success and failure dynamics as we look at them through the lens of history. What has failed as an ingredient of an old era model may succeed in the model of the future. Conversely the success criteria of the old may represent the failure criteria of the new. We need to look at everything through the lens of our intended destination and how our mindsets and models need to adapt to the cultural and geo-political context of that future.

Another economist who is evolving mindsets in our desire to optimise financial models is Mariana Mazzucato. In her book, *The Value of Everything*, Mazzucato explores how the prevailing societal mindset finds value extraction more rewarding than value creation. Taking out rather than putting into the system has been encouraged by the deregulation of the financial sector, extracting value from sectors which they had used to mobilise resources to create value. Often this is fuelled by the drive to maximise (often short-term) gain for shareholders. Value creation can take longer than the value extraction timelines of shareholders. By placing success and reward in a place of short-term value extraction we encourage unproductive activities and neglect productive forms of value creation. She concludes that value depends on vision: 'If we cannot dream of a better future and try to make it happen, there is no real reason why we should care about value' (Mazzucato, 2019, p. 280).

Mazzucato also respects the role of governments in setting the platform for the corporate mindset to think differently by the encouragement of alternative economic models and mindsets. In *Mission Economy* (2021), Mazzucato points to the dysfunctional form of contemporary capitalism, fuelled by and fuelling the climate crisis and a practical framework for ambitious, state-led climate strategy.

Adapting Beyond the Boundaries

Before exploring the characteristics of old and new era capitalism, a quick diversion into some basic principles of Darwinism.

Darwin's concept of 'survival of the fittest' is often seen as being at the root of the prevailing model of capitalism. This refers to the deterministic philosophy of Englishman Herbert Spencer, who applied Darwinian biological and evolutionary concepts of natural selection to humans and markets. Spencer argued that 'human progress resulted from the triumph of superior individuals and cultures over their inferior competitors; poverty was evidence of inferiority' (Young, 2012).

To suggest that we can capture Darwin's learning of natural systems and apply them in such a crude and one-dimensional way completely misses the opportunity that nature presents to teach us how to thrive indefinitely in a system with abundant (but finite) natural resources and no waste.

In *The Origin of Species,* published in 1864, Darwin shared the research that underpinned much of his work on the Galapagos Islands. Perhaps the best known of the species he collected while on the Galapagos Islands were what are now called 'Darwin's Finches'. In 1836, while back in Europe, Darwin enlisted the help of John Gould, a celebrated ornithologist. Gould was surprised to see the differences in the beaks of the finches collected and concluded that they were in fact different species. Moreover, 12 of the 14 finch species were deemed to be brand new species.

It was this discovery that solidified Darwin's thinking around natural selection. He observed the favourable adaptations in the beaks of finches over generations until they branched out to make new species.

Why was this so important? The finches adapted to the types of food available to them when they arrived from the mainland to the Galapagos. This allowed them to fill different niches in their new environment. By developing beaks that would give them access to new food sources, it allowed the population of finches to expand within the ecological boundaries within which they found themselves. It was this discovery that allowed for the discussion around how species change over time, including divergent evolution, or adaptive radiation – a term used to describe a rapid increase in the number of species with a common ancestor, characterised by great ecological and morphological diversity.

So, what is the relevance of this detour into nineteenth-century Darwinism?

The finches teach us about the importance of adaptability. They recognised the boundaries of their operational environment. They adapted to maximise growth and to thrive within their context. What they didn't do is seek efficiency improvements, or exploit opportunities to outcompete each other for the same food source. There will have been some superior performance and thriving of the fittest birds in this system, but to call that one component out of the findings of Darwin misrepresents the learnings we can take from these studies into the design of the systems for the new era.

The world provides us with abundance, but we contain ourselves by operating in a system that focuses on outcompeting for the same food source. Operating within planetary constraints sounds restrictive, and it does apply some guard rails, but through changing our mindset we can find more abundance within our planetary boundaries. Abundance is related to capacity. The carrying capacity of

planet earth is determined by the value we place on protecting and conserving the assets of nature. If we destroy
nature then we reduce the carrying capacity and we reduce
abundance. So our current model of capitalism fails us on
two counts. Firstly, it constrains us in our ability to unlock
the abundance that is available. Secondly, because nature is
not valued and is therefore exploited and destroyed, this
reduces the abundance that is available for us to unlock.

The current system is so prevalent that whatever type
of organisation you find yourself in, you are likely to be
faced with the question of how you win over the competition, how do you beat them? This is often individualistic
in focus. You are steered towards the concept of your win
often equating to your competitor's loss. This is a zero-
sum game. Let's reframe the question. How can your community thrive by adapting within the finite boundaries
that govern your existence? Competition will continue to
be a human characteristic that will bring about progress,
but instead of fighting over the same food source in a
zero-sum game, the reframing of the question should drive
us into an uncontested space that calls on the other human
qualities of co-operation and nurturing.

Recalling the ventilator project that we shared in
Chapter 1, the participants set about their task by putting
the competitive forces of the old era to one side. They
asked the question posed earlier: How do we, as automotive and aircraft manufacturers, adapt what we do in a
pre-competitive, collaborative space to deliver superior
results? How can we call upon nurturing instincts (as
important a human behaviour as competitiveness) to
mobilise our energies into a process where we can stop 1000
people dying per day through collective adaptation?

As we shall see, adaptability will become a critical mindset, an asset that will be required for success in Day 2.

New Business Models

We discussed the adaptability mindset at work in Nestlé Nespresso's coffee business model with CEO, Guillaume Le Cunff.

> *At Nespresso we operate a 'Direct to' business model that embraces full value chain integration. That sees the value chain as an ecosystem with all stakeholders being equally important. An ecosystem where customers can connect to farmers. It is these connections between our stakeholders, between farmers and customers, that makes Nespresso a real luxury brand.*
>
> Le Cunff, 2022

Le Cunff likens the Nespresso business model to a top chef: 'They will tell you about where the food is from, the way the animals have been kept, their diet, or how fresh the vegetables are and in what conditions they've grown.' Similarly, the authentic and personal approach, knowing all the details of your raw materials, the people that work tirelessly to produce these, and each step of the process is all part of the story and the brand appeal. He describes it as 'a passion and a duty. If you want to be a luxury brand, you need to know every step of your value chain and connect directly with each and every stakeholder in this value chain.'

The business models we choose for the new era will need to demonstrate financial performance to thrive. But they will equally need to perform in the context of their entire ecosystem of stakeholders inherent in the nature of stakeholder capitalism. As Mazzucato says, 'The stakeholder theory of business is more than a theory of how to run a company better; it also has far-reaching economic and social implications. It answers the question, "What makes a business successful?"' (Mazzucato, 2019, p. 184).

Guillaume Le Cunff is pursuing such a balance, driving Nespresso to meet ambitious sustainability goals. This is captured most pertinently in their recent B Corp certification. B Corp is the international network of organisations driving systems change and a vision of an inclusive, equitable, and regenerative economy. B Corp was originally a network of smaller, purpose-driven businesses seeking differentiation in congested markets through their public commitment to all stakeholders, but Nespresso's certification marks a shift. The balance is fine. A business like Nespresso will struggle in the transition to immediately embrace total transformation and remain profitable, but Nespresso's posture is facing in the right direction, and the leadership is committed to a sustainable future and with a clear vision to embrace business models that fully reflect this.

The B Corp movement and vision which drives this, contrasts with traditional shareholder responsibility models where a board of directors is elected by the shareholders to govern the management team and to make corporate decisions on their behalf. The directors have a fiduciary duty to protect and manage stakeholders' interests in the company.

We are not proposing that we lose sight of the importance in the relationship between the directors and those who have taken a leap of faith to invest in them as a management team and the business they represent. Our view is simply that by shaking loose of this one-dimensional obsession, and by engaging with all stakeholders within a Total Value System, decisions are more likely to be taken in line with the drivers of business success. Ultimately within a Total Value System integration model, all stakeholders should gain. This includes delivering superior long-term returns for the shareholders.

Allison Dring is CEO and co-founder of Made of Air, a cutting-edge business making carbon-negative materials to transform products in mobility, consumer goods, and the built environment. For Made of Air, planet Earth is a key stakeholder in the business and 'taking care of that relationship is as important as the other stakeholders in our business':

> *Contrary perhaps to expectation, this clause, which even in a climate-friendly business is bound to come into conflict with economic decisions, not only didn't put off potential investors, but for some it was the part that sealed the deal.*
>
> *When the WEF stated in 2021 that the future of the economy depended on sustainability and that the money to be made in the future would be tied to sustainable measures or technologies, I think that really resonated with a lot of investors we were talking to. A lot of them could see that if you don't get into this space and start investing in this technology then you'll lose out. We don't have other options.*
>
> Dring, 2022

Delving further, we can see how embracing all stakeholders in a Total Value System might start to unsettle the given norms in how business drives value. Porter's Five Forces model is widely used to shape most industry sectors and segments of the economy to determine weaknesses and strengths against five competitive forces that shape every industry and every market:

1. Competition in the industry
2. Potential of new entrants into the industry
3. Power of suppliers
4. Power of customers
5. Threat of substitute products. (Porter, 1980)

Within a model of stakeholder capitalism, the success of business enterprises operating in free markets will still find relevance in this Five Forces model. But we are starting to see divergence from it and signals of the forces that are promoting the new business models for the new era.

1. **Competition**
    The zero-sum game form of competition, where one company's win equates to another company's loss. The ventilator example or the open sourcing of medical data for the mass production of vaccinations shows that collaboration provides us with enlarged pre-competitive spaces.
2. **Supplier loyalty**
    The retreat from the over-reliance on buyer leverage over supplier leverage. In the current model the sourcing of inputs is typically all about cost. Value is derived from negotiating cheap prices and maximising this arbitrage with your sale price to customers. Instead, as with Nespresso, we see a move from a supplier leverage model to a supplier loyalty model where supply chains are held differently in the value chain. By moving to an inclusive rather than an exploitational model, the buyer will enjoy stickier relationships, greater security of

supply and a stronger story to tell customers against a backdrop of increasing transparency.

3. **Customer Empowerment**

We are moving from a linear to a Total Value System: from the 'Power of Customers' to the 'Empowerment of Customers'. As with Tony's Chocolonely, we expect to see a greater sense of identification by customers with a brand that is less functional, and more heartfelt, that connects emotionally in an era of transparency and authenticity. As a chef derives value from understanding the origin of the food served, so too we'll see customers placing greater value on products and brands that are able to operate with greater visibility across value chains. This will empower customers to connect with all parts of the value chain and make wise and values-driven buying choices. To a customer this looks like enhanced value, and this, especially in uncongested market spaces, would typically lead to improved margins.

4. **Closing the loop**

Relationships with customers will reflect changing buying habits but there will also be a functional relationship driven by Day 1 compliance needs. As we shall go on to see, economic models are moving from linear to circular. Your relationship with clients needs to change to allow you to track and retrieve your product from them for recycling and reassembly in your supply chains. Failure to do this may lead to the threat of disruption from a company that is more effective at operating in the circular economy than you, creating substitutes and selling your product, reassembled, for a cheaper price.

5. **Social impact engagement**

New business models introduce specific nuances that require the deep knowledge and understanding of their societal context. Partnering with Non-Governmental Organisations (NGO's), for example, can give opportunities for brands to collaborate meaningfully using organisations that are engaging with government and civil society. Canoeing in Day 2 involves engaging wisely with regional and local society and exploring ways by which a company demonstrates how it addresses systemic social and economic injustices within its business model. New business models need to be bold to classify what is within and without their scope of social engagement.

Most businesses operate in a model where the primary activities consist of inbound and outbound logistics, marketing and sales, a service department, and operations. These are supported by human resources, technology, and procurement. This is the corporate stronghold.

The supply chain is outside the stronghold. Value is created from leveraging as much as possible from suppliers with the procurement department defending the walls of the stronghold. Operations convert cheaply sourced raw materials from the supply chain into a finished product or service, effectively changing all inputs and readying them as outputs. As a product becomes outbound from the stronghold, the sales team defends the walls at the point of sale. The customers are also outside the stronghold and producers are not responsible for the end of life of the products they release to customers.

As we start to compare and contrast the divergent practices identified above, we observe the move away from the 'norms' of business as we know it. These are identifiers of the shifting and crumbling of strongholds as we move into a new era.

Circular Value Chains

As touched on above, perhaps the greatest disruptor of traditional value chain thinking is the move from linear to circular business models.

A *circular economy* or business model is a model of production and consumption which involves sharing, leasing, reusing, repairing, refurbishing, and recycling existing materials and products with the aim of eradicating waste. It has become clear that for us to meet our climate targets we need to move from the wasteful model of the linear economy to a circular one. But going circular is not just about climate impact and the associated use of energy in the linear economy, it's also about biodiversity loss, pollution, and the suffocation of our oceans. According to the Ellen MacArthur Foundation, there are three primary principles required for the transformation to a circular economy:

- eliminating waste and pollution
- circulating products and materials throughout a circular value chain
- the regeneration of nature.

This is an area where the Charge of Day 1 is particularly threatening. The compliance requirements are reflective of an increasingly burdensome regulatory environment. Whether it's the Circular Economy Action Plan, the European Green Deal, the Plastics Pact in the UK or the EPA National Recycling Strategy in the United States, we should expect compliance to break the strongholds and plunge many businesses into the transition zone.

But there are businesses that have used the vision of the circular economy to launch themselves into the opportunity of Day 2. Rapanui is one of these.

Mart Drake-Knight, the founder of circular clothing company Rapanui, explains how they wanted to explore closing the loop on an industry where 60% of clothing is made from plastic, 99% of clothing waste is landfilled, and 40% over-production occurs as clothing is made on expectations of demand rather than actual demand. They discovered the reason clothes are not made to order is because 'it would mean producing things one at a time, in real time, and at a speed and scale that would require hundreds of people looking after thousands of processes, making millions of decisions round the clock, 24/7, almost at the speed of light'. However, they realised that while the human brain can't process that much information, computers can. 'And when we made that connection, it was something we acted on' (Drake-Knight, 2022).

Rapanui makes clothes to order, seconds after they are ordered. They save so much money on waste reduction that they are able to source better organic materials and focus on the production of a better product, alongside passing some of the savings on to the customer, creating a better value product. Additionally their clothing contains a QR code that can be scanned for easy returns at end of life, designing in a fully circular model and, with the help of their customers in their reverse logistics, they offer discounts on future purchases, engendering customer loyalty.

Rapanui are redesigning the business model by bringing the Total Value System into their primary activities, by finding uncontested space and out-competing clothing companies that are seeking 'performance' by embracing

Sustainable Performance. They have 'reapproached' and provide us with a glimpse of the future.

Most businesses face a challenge of transitioning from the compliance of Day 1 to the opportunity of Day 2. This transition, as we have described, will require stepping out of strongholds and into danger zones that will expose the value of courageous leadership. To make this transition easier there are companies that provide reverse logistics to help close the loop for organisations that remain in the forward logistics 'take, make and throw away' model of the linear system.

Bruce Bratley, CEO of First Mile, a leading recycling business based in the UK, describes the challenge of operating within the prevailing system:

> *I see the current market as super highways sending products and services in one direction. Let's say there are 10 million types of businesses in the world producing products and services that they're sending down those superhighways. But there is one sector, which is the waste management and recycling sector, that is trying to capture the waste at the end of the road. . .you just get all of the waste product with no data, and then you're expected to deal with it and somehow send it back out into the reverse supply chain.*
> Bratley, 2022

Bratley acknowledges the enormity of the challenge but is clear that businesses ultimately have little choice but to change.

> *They have to do something because the current model isn't going to exist in 10–20 years. As a business leader you're going to be saying, 'I've got climate risks, I've got supply chain risks, I've got human capital risks,' they all point to the same thing, which is we've just been extracting too much too cheaply.*

Rather than having separate sectors doing different things we need to have all of those businesses looking at their resource base, their consumption, and saying, 'How can I turn my business from being a consumer-based linear flowing business into a regenerative business which can take back materials, can backup reserve materials, get our own plastics back and turn them back into new products or turn them into products in a parallel sector?' Everything's going to have to be reusable and everything with built-in obsolescence needs to be designed out.

Nature and Food Supply

If there was one sector that we would expect to be an honouring partner to nature, it's farming. Sadly though, farming has not been immune to the world's insatiable appetite for taming the natural world. This has been exacerbated by farming typically being outside the stronghold. Operating outside the stronghold in the supply chain usually means less leverage than the typical buyer leverage within the stronghold. The demand from within the stronghold of the business, that is then passed onto the farmers, is more yield for lower cost. This may mean more fertiliser, more pesticides, more automation, leading to economic and social division. The pressure is to provide more, at a quicker pace, for lower costs.

René Toet, Managing Director of Climate Neutral Group, is a pioneer in the reconnection of capital markets with nature and in mitigating climate impact. René describes one project in South Africa:

We want to use carbon finance to improve land management practices. In South Africa the practices of dairy farming have typically stripped 60 per cent of the nutrients from the soil. We want to support the farmers to move from this degeneration model to a regeneration model. We focus on mainly five sources of regeneration:

1. *Using less fertiliser*
2. *Improving water management*
3. *Improving grazing practices*
4. *Reducing tillage*
5. *Improving crop planting*

By focusing on these five we'll see soil quality improve, we'll use less carbon and we'll create the right soil conditions for it to act as a carbon sink. If you operate in an approved methodology framework, then these carbon savings can convert into credits with an economic value that can be sold to help organisations address their residual emissions and meet their net-zero targets. While these credits can be sold in South Africa, they will be internationally certified and so could be sold to any company around the world. Most of the value of these credits go to the farmer. We plan to get 750 farms signed up in South Africa and we are now taking the same projects into South America.

Toet, 2022

In a traditional model the buyer gains leverage from holding a supplier in a place of dependency and then uses this leverage to drive down the cost, seeking competitive alternatives to further reinforce this leverage. The farmer has their own needs to protect. So, the only way to manage the farmer's interests, and the buyer's stronghold, is to find a method of agriculture that further enhances yield at a lower cost.

Arguably no stakeholders benefit from this short termism, and a model that is exposed to vulnerability and unquantifiable risk operating outside planetary boundaries will ultimately destroy value. But let's call out the stakeholders that are both losers and completely neglected in this model – nature and future generations. Sixty per cent of the value of nature is stripped in the example above. The only way to support this broken system is to use more fertilisers and pesticides.

This vicious cycle driven by the current short-term economic model is now breaking. Supply chain vulnerabilities are many and varied – from conflicts such as the one in Ukraine, and logistical failures in the distribution of previously available inputs such as fertilisers or red diesel for machinery, to the costs of energy and extreme climatic events. These all expose the vulnerability of a model that controls or neglects nature as a key stakeholder.

Perhaps it is not surprising that in parts of the world farmers are having to employ people with paint brushes to manually pollinate the crops because the bees and small insects that nature provides us to do this work have been killed by excessive pesticide use.

The South African farmers in the example shared here are moving away from the complex and ultimately destructive nature of their Day 1 practices to a more harmonious Day 2. They are reapproaching, with nature as a stakeholder. They are sourcing alternative forms of capital within this reapproach. They are reducing their dependency on those downstream to them, thus readdressing the power that farmers should have in the value chain. They are providing the value chain with lower carbon alternatives and customers with products that reflect their changing buying habits. They are attracting more talent as they become more purpose-driven and more prosperous to provide improved rewards. Young people are staying on the farm rather than putting migratory pressure on nations and governments. The social cohesion of the community remains intact. Future generations have hope.

These are powerful new ideas that carry the expectations of the emerging generations who know that their future depends on us getting this right. We shall not deliver the 'pivot and perform' transition that will be required by tinkering around the edges. Flaws in the models of the old era are being exposed. Change is inevitable and we face the decision to build the walls higher and stronger to extend our increasingly challenging existence, or to transition through the danger zone to prosper in a redesigned world of the next era.

These traditional institutional structures are being shaken as the regulators, customers, employees, and investors are, in different ways, calling for the Total Value System to be within the business ecosystem. All stakeholders, including planet Earth, have a stake and a voice in the decisions made. Change is ultimately being enforced because the sustainability targets that are being set and agreed to, can only be met through a Total Value System integration model. The push-back of nature is accelerating the process. We are learning that externalities related to nature cannot be tamed in the way we thought. We must work with nature in a more adaptable way and find value chain models that build resilience in the face of the Charge from increasingly extreme climatic and nature-based interventions.

Klaus Schwab, founder and Executive Chairman of the World Economic Forum, describes the challenge of moving

from a shareholder-focused system to a stakeholder model
of capitalism and reimagining our global economy so it
becomes more sustainable and prosperous for all:

> *Our global economic system is broken. But we can replace
> the current picture of global upheaval, unsustainability,
> and uncertainty with one of an economy that works for all
> people, and the planet. First, we must eliminate rising
> income inequality within societies where productivity and
> wage growth has slowed. Second, we must reduce the
> dampening effect of monopoly market power wielded by
> large corporations on innovation and productivity gains.
> And finally, the short-sighted exploitation of natural
> resources that is corroding the environment and affecting
> the lives of many for the worse must end.*
>
> Schwab, 2021

There is much to take on board here, so for those
looking to embrace the power of stakeholder capitalism,
we have summarised the primary components at the end
of this chapter.

Here we have framed the 'Why'. Fundamental
changes to the economic, capitalist, and business models
to which we have become accustomed are going to be
necessary and there will be major implications for every
member of your leadership team. This may appear dangerous, massively uncertain, and hugely complex. It will
represent change and there will be decisions to take at
every level, but with the right leaders and guides, the
opportunities are abundant.

In the chapters that follow we will remove some of the
complexity and seek to bring clarity as we guide you
towards the process of 'How'.

How to Embrace the Power of Stakeholder Capitalism

- Look to bring the entire value chain into a redesigned model where all stakeholders can benefit.
- See your value chain as less of a chain and more of an ecosystem.
- Move from a model of buyer leverage over suppliers and seller leverage over customers to a model of supplier and customer loyalty.
- Bring your staff into the equity wealth creation. Consider extending this to suppliers and customers.
- Break down the walls of the existing stronghold and save the costs of defending it. Extend the walls to the edges of the value chain and engage the resultant ecosystem in designing for reuse, bringing long-term resilience to your product or service by closing the loop on your supply chain.
- Mobilise your fully integrated value chain to become better at reusing, repurposing, and recycling your goods and services than those companies who may seek to take your market share by reselling your goods and services.
- Adaptability over efficiency, collaboration over individualism. Compete AND nurture.
- Define your finite boundaries and adapt to find uncontested space.
- Work with nature. It's your partner. There may be a cost, but it will deliver abundant returns.
- Extend the time horizons of your shareholder expectations.

References

Bratley, B. (8 August 2022) Interview with the authors.
Drake-Knight, M. (31 March 2022) Dear Mr Bin Man. TED talk. Retrieved 7 December 2022, from https://www.ted.com/talks/ mart_drake_knight_dear_mr_bin_man?language=en

Dring, A. (5 September 2022) Interview with the authors.

Ellen MacArthur Foundation (n.d.) What is a circular economy? Retrieved 7 December 2022, from https://ellenmacarthurfoundation. org/topics/circular-economy-introduction/overview

Le Cunff, G. (10 August 2022) Interview with the authors.

Mazzucato, M. (2019). The Value of Everything: Making and Taking in the Global Economy. Penguin Books.

Mazzucato, M. (2021). Mission Economy: A Moonshot Guide to Changing Capitalism. Penguin.

Porter, M.E. (1980) Competitive Strategy: Techniques for Analyzing Industries and Competitors. Free Press.

Raworth, K. (2017) Doughnut Economics: Seven Ways to Think Like a 21st Century Economist. Random House Business.

Rockström, J. (2015) Big World, Small Planet: Abundance Within Planetary Boundaries. Yale University Press.

Schwab K. (2021) Stakeholder Capitalism: A Global Economy that Works for Progress, People and Planet. Wiley.

Toet, R. (5 July 2022) Interview with the authors.

Young, W.H. (19 April 2012) 'Capitalism and Western civilization: Social Darwinism. National Association of Scholars. Retrieved 7 December 2022, from https://www.nas.org/blogs/article/capitalism_ and_western_civilization_social_darwinism

6

Leadership
Leaders, Organisational Will, and Guides

In this chapter we'll explore the call on the leaders of today and tomorrow, of Day 1 and Day 2, considering those focused largely on managing and shoring up the stronghold versus those leading away from the Charge through the transition into Day 2. We will also see the emergence of purpose-driven entrepreneurs motivated by the challenge of our time. Without the clutter of a Day 1 legacy, they will jump straight into Day 2, propelled by the accelerators and incubators seeing opportunity in the next industrial revolution.

Leading Beyond Compliance

We have set out throughout this book the burgeoning nature of the general ESG compliance agenda and the avalanche of new climate and sustainability related legislation in multiple jurisdictions. The cost and operational burden of all of this compliance activity is only going to get greater.

This can lead to frustration and some leaders may even ask themselves how this agenda aligns with why they aspired to organisational leadership in the first place. They may question their fundamental motivation – is it to manage compliance, survive the next reporting cycle and do it all over again in the following reporting period? Is it to look at the minimum additional investment required to stay in the game and report some marginal improvement against indicators that they see as peripheral to their organisational purpose? Another set of leaders might naturally prefer to make the transition to Day 2, but the incessant noise, the complexity and confusion of the compliance and regulation agenda might be driving frustration, inaction, and stagnation, not innovation and enterprise.

Of course, other leaders may in fact find this management of compliance very motivating. They might pride themselves on their ability to secure a licence to operate and survive year after year without having to consider anything more substantial or disruptive – just remaining at the helm of the organisation and keeping it sailing steady as she goes. Such managers might see other spheres of organisational activity as the areas to invest in and be disruptive in their respective industries. Inspired by some of the 'Blue Ocean', 'Infinite Games' 'Competing for the Future' type strategies, they might see other category drivers as the opportunities for game-changing ideas that reinvent and create new sources of value. For them sustainability will remain just a dashboard of compliance Key Performance Indicators (KPIs) that misdirects them to do the stuff that they perceive drives value in their business.

Then there are the leaders that need no convincing. They have already been drawn into leadership excited by the opportunities of 'enterprise'. Of course, they well understand the need to continue to secure the licence to operate of their respective organisations, but they realise early on that compliance for compliance's sake is insufficient motivation for themselves or their stakeholders.

Such leaders know implicitly that the purpose of enterprise will always be higher or greater than compliance. For them the compliance agenda lags their ambition for performance. Guillaume Le Cunff, the CEO of coffee company Nestlé Nespresso, framed it this way: 'We want to comply, and will. But a new compliance cost can trigger a source of value and ultimately it will open up new wider opportunities for us to innovate and grow' (Le Cunff, 2022).

Such leaders understand deeply that the organisational assets that have been entrusted to them (their brands, supplier relationships, customer base, investors, intellectual property, and fixed assets) are valuable, powerful, and can be leveraged to higher effect in the wider world than simply ensuring they are being managed in line with any set of rules and regulations. They can be unlocked as sources of new value creation that go beyond a compliance agenda. Leaders with this perspective see beyond the rules and understand the more foundational rationales behind the compliance – and here they see opportunity.

We'll use an example – the story of the talents. A talent was a currency in the Ancient Middle East. The story goes that a landowner and lord was going away on a journey and called his household servants to him. He entrusted each of them with a number of talents. One received five, one received two, and the last one just one. He asked his servants to take care of these talents and use them wisely. On his return he called the servants to him and asked them to make an account of what they had done with their talents. The first had invested five and was able to return 10 talents, the second had also invested and returned four. The last servant said: 'Master, I knew you to be a hard man, so I was afraid, and I went and hid your talent in the ground. Here, you have what is yours' (Matthew 25:14–30, The Bible, English Standard Version, 2001). Assets are talents and we can bury them in compliance and fear or invest them in the wider world to deliver a multiple and abundant return.

For others though, they will find intrigue in the prospect of a Day 1 to Day 2 adventure. It will prod at their pioneering instinct, and they will find a charismatic way to bring their ideas to life. Some of these never really set out to be business leaders in the first place, at least not in the sense of aspiring to run large organisations. They had other motivations beyond leadership. These are people with curious minds, who are imaginative, perceptive and sensitive. They are people interested in the world and how it works. They are pioneers with energy and enthusiasm. Some of them are scientists, others are activists, all of them thinkers. Whether spiritual or secular, they care about humanity and the environment.

The eponymous founders of Ben and Jerry's identified themselves as two hippies from Vermont. In their seminal book *Double Dip: How to Run a Values Led Business and Make Money Too* (Cohen, Greenfield, and Maran, 1998) – which, incidentally, was one of the stories that set Dean on his sustainability career – they talk about how they wanted their business, first and foremost, to have an impact.

> *We love making ice cream—but using our business to make the world a better place gives our work its meaning. Guided by our Core Values, we seek in all we do, at every level of our business, to advance human rights and dignity, support social and economic justice for historically marginalised communities, and protect and restore the Earth's natural systems. In other words: we use ice cream to change the world.*
> Ben and Jerry's, 2022

Ice cream was just a vehicle through which and by which they could generate that impact. In the book we learn that it could just as easily have been pizza.

It is interesting to observe how a brand like Ben and Jerry's has actually inverted the strongholds of consumption and consumerism, and consumer product marketing dynamics and the power of brands to create impact and value. It's also revealing to see how consumers have chosen to join them in that endeavour.

As Ben and Jerry tell their story, time and again it becomes apparent how they think differently. Their progressive social agenda is what motivates them. Even though the company is for profit, it has enjoyed tremendous success in the marketplace and its products are high-quality and premium. They set out to run a values-led business and the 'make money too' always seems not quite an afterthought, but an outcome rather than an objective. *Double Dip* remains one of the most simultaneously imaginative and practical accounts of Sustainable Performance in action.

Unsurprisingly unusual leaders do unusual things and Day 2 therefore feels to be more familiar territory to them.

In our discussions with Dame Polly Courtice, who led the Cambridge Institute for Sustainability Leadership (CISL) and sensitised, trained, and equipped thousands of leaders in sustainability over 32 years, we (both Stuart and Dean) were flattered to be called 'unusual consultants'. What she meant was that it was obvious our primary motivation is not just peddling sustainability consultancy services but co-exploring and guiding our client partners on a journey towards sustainability-inspired value creation and Sustainable Performance – we help our clients to uncover solutions which are unexpected, unusual, disruptive and differentiated; ideas that are sustainable, innovative and return to the spirit of true enterprise.

Whatever the motivation of the leader – trying to manage the demands of Day 1, managing Day 1 while exploring Day 2 or being essentially driven by Day 2 opportunity, the reality is that the binary choice in sustainability is no longer a choice open to any of these leaders. It is increasingly clear that circumstances have made the choice for us. The reality is that all organisations will need to comply, but compliance alone will no longer guarantee survival. In an email discussion in 2022 with Tom Williams, head of Nature at the World Business Council for Sustainable Development, he referred to the framing of the net-zero and biodiversity challenges as 'beyond value chain'. That means that even the aggregation of the positive outcomes of all the Science Based Targets and the net-zero ambitions of organisations will not offer the speed or the scale to address the issue. The rate of change is such that all organisations will inevitably be forced towards a predominantly Day 2 agenda and here are the reasons why.

It is our conviction that whether you already plan to stay firmly rooted in the *terra firma* of Day 1 or if Day 1 is your natural default, then even hunkering down against the Charge and manning the barricades of the stronghold will only save you in the short term. True, that might be enough for your tenure or to secure your pension, but it will not build the lasting resilience the organisation needs, and investors and other stakeholders will be wary of leaders with this limited view. Leaders with a compliance-only leadership mindset and strategy ultimately view this as an endurance test and believe their strongholds will prevail if they can only defend them long enough. We are already well and truly in the danger zone and the strongholds grow weaker by the day, as we saw in Chapter 2.

As we recognise that an infinite model relying on finite resources is fundamentally a broken system, we become lured by the canoe for a reapproach that will provide us with the reliability and resilience of Sustainable Performance.

The third part of this book moves into the adventure of Day 2, the Reapproach. How can we take control in this time of disruption? Or perhaps more realistically, how do we set a different frame around what we do and shift from command and control' to models of 'empowerment and trust'? We are moving from managing the risk of the Charge to the opportunity of the Reapproach.

Day 1 to Day 2 – Leading Through the Transition

In the late 1700s, William Wilberforce, alongside other members of a group of social reformers called the Clapham Sect, were in a liminal space.

The emotional climate around the use of slaves to stimulate growth and wealth in the economies of rich nations was being questioned. Slavery had been part of the system since the fifteenth century. Nations, economies and the private wealth of individuals and families were dependent upon this system to service the storehouses within the strongholds of the prevailing era. Between the sixteenth and nineteenth centuries, approximately 12 million Africans were enslaved and transported to the Americas. As moral disapproval grew, so the cracks started to appear in the stronghold.

As we've highlighted, when strongholds are attacked, anxiety is quelled by extending the truth into fantasy. Stories were propagated that enslavement was good for African

Americans, that slaves were content under the care of their masters. Comparisons were made between the relatively lucky slaves on the plantations and the factory-based slaves found in other strongholds. Underlying these arguments was a belief system in the superiority of whites and an economic dependency on the wealth created by the system of slavery.

The central character in this story was William Wilberforce, becoming an independent member of the UK Parliament in 1780. Wilberforce surrounded himself with powerful social reformers driven by their faith, by education, a passion for truth, by a moral conviction, and a burning desire to fight the horrors of the slave trade. Motivation can be a complex issue but the 'Claphamites' had a purpose.

By 1807 Britain passed the Abolition of the Slave Trade Act, outlawing the British Atlantic slave trade, and ordering gradual abolition of slavery in all British colonies. Most of the other economies around the world that were active in the slave trade followed soon after. As we have seen at liminal times in history, change happens slowly and then all at once.

It seems that when Wilberforce and the Clapham Sect presented their Day 2 scenario, they met fierce resistance, and yet a small group of people were able to transition the liminal space in a relatively short period of time and influence the design of the new era.

Wilberforce was said to lead through a character of humility, trust, honesty, and care. He had the qualities of an ethical leader and he earned the respect of others, asserting an empowering rather than controlling leadership style. He was courageous and was willing to step into the danger zone by partnering with a strong moral conviction. Instead of rampaging through parliament engaging with verbal attacks against his opponents, he would seek to build trust and unity across political divides. We will come back to this but for now, back to our liminal space in 2023.

Let's not be under any illusion that leading through deep complexity and uncertainty to an unknown destination is going to be hard or easy. The type of leadership required to navigate shifting strongholds, liminal spaces, old eras moving to new eras, complex societal challenges, alongside competing expectations from all sides of your business, will require different skills to that of just managing the stronghold.

Wilberforce was a transitional leader. He and the Clapham Sect needed to understand the stronghold, the competing pressures and weight of expectation on the leaders of their Day 1. They needed to understand anxiety and fear and the risk of leaders seeking an 'off ramp' into a comfort zone facilitated by the creation of the fantasy zone. With empathetic but courageous leadership they reached out to leaders and they guided them into the transition zone by seeking unity over combat.

Our friend and storytelling coach, Stew Bewley, CEO of Amplify, teaches us about the hero's journey:

> *Most people stare into the hero's journey most of their lives – habits they are stuck in, relationships and jobs they can't escape, desires they never see fulfilled, the monotony of existence. But then comes what screenwriters call an 'inciting incident' – something that tips the hero over the edge into their journey. In the movies it will be an invitation, from Gandalf to Frodo to go on the journey of destroying the ring, or a choice, from Katniss Everdeen to take her sister's place in the Hunger games, or an event where the character is forced to make a choice, like Andy Dufresne in the Shawshank*

Redemption getting sent to prison – will he persevere, survive and thrive in prison or give in to its oppression? Everyone has to make a choice with what life offers us, sometimes it offers us adventure (Lord of the Rings), sometimes it offers us suffering (Shawshank Redemption).

Who we are in that moment and the choices we make, is when we truly embark on the hero's journey. That is when life gets interesting – the twists, the turns, the pitfalls, the wise mentors along the way, the treasure we find at cost to ourselves, the dragons we kill, the new land we climb back up to, changed by our experience, with our scars, but stronger and more resilient. Those are the stories we need to be telling ourselves, our friends, our work colleagues, as often as we can. Why? There are so many distractions in life to shatter our stories to pieces – the instagram post that sinks us into comparison, the news report that sinks us into despair or propels us to moral outrage. But all of these stories don't originate with us. They are forced upon us and drown out the anchoring truth of the actual story we are living.

Bewley, 2022

Transitional leaders recognise that the strongholds of the old era, which have been foundational in their success, security, and wealth creation, are over. These leaders are embarking on a hero's journey. Their story will involve high individual and organisational effort in Day 1 to stay in the game with tenacity, innovation, and courage to thrive in Day 2. They will need to be tactically skilful, humble in the way they acquire knowledge and wise in the way they use it. Their destination is 'Sustainable Performance' and en route to the summit they will need to have conquered the peaks of net-zero, circularity, inclusion, and stakeholder capitalism.

Transitional leaders will need to lead from conviction, have a clear and motivational purpose. Purpose without authenticity is nothing and leaders that lack integrity and the ability to operate in the truth will likely fail early in the journey. The heroes of the Day 1 to Day 2 transition will respect the need for guardians to steward the compliance requirements of Day 1 but they will choose another team of pioneers to join them on Day 2.

You may have faced an 'inciting incident' in business or home life that challenges your mindset, behaviour, or career path. The business challenges could be constraints on capital, loss of talent or market share, the collapse of supply chains, or escalating input costs from climate impact.

For many leaders it is not about a sudden eureka moment of conversion, but a more gradual realisation of the need, and the opportunity to think differently about their business. For Sebastian Heinzel, the CEO of Heinzel, a global pulp and paper company, it was a process:

> *I have always cared about the big picture – I didn't want to go into the family business – I was more interested in history and philosophy. When I came to the family business, I was always clear I did not want to just sail it in the same direction. When I became a substantial shareholder it triggered that thought process and then I spoke with friends and I knew I was committed to this agenda and this path. Having kids has also shaped my perspective. When I thought about that I simply asked myself – what do I do with this asset?*
>
> Heinzel, 2022

But challenges can come in very unexpected forms: a child returning from school asking about how you are using your leadership role to safeguard their futures; or, a development in your moral convictions. A recent conversation with Chris Iggo, Chair of the AXA IM Investment Institute and Chief Investment Officer of AXA IM Core captured the essence of the dilemmas leaders have often faced and the way in which a reapproach can help to resolve them and create opportunities for both professional and personal growth.

I grew up in nature, I love nature. But I was good at maths and the natural passage for me was to go to a top university and then into the City. I followed the rules and did well. The system allowed me to earn a good salary, but deep down I knew I was working in a system that was not valuing something that was more valuable than any of the material gain that formed the basis of my reward. Philanthropy didn't do it. My wealth was created in ignorance of its impact on nature or at the expense of nature and I couldn't adequately compensate for this by giving back as an individual. I wasn't partnering with something I loved, I was part of exploiting it for my own gain. A gain that, in theory, should be enjoyed by me and my family but increasingly my children were challenging me on this and suggesting it is a flawed ideology.

Iggo, 2023

In our interview with Gina Mastantuono, CFO of software giant ServiceNow, we asked if there was a light-bulb moment that triggered her passion for sustainability. She told us that as a mother of two, the light bulb moment continues to burn. I can't remember if there's been an 'aha' moment, but my concern and determination to leave a better earth for my children and loved ones, and future generations is something I'm very passionate about. I'm also thrilled that I can channel this passion through my work at ServiceNow where I lead our ESG strategy. (Mastantuono, 2022).

These light-bulb moments of enlightenment and insight may cause us to see where we have succumbed to strongholds without a fight. We begin to ask ourselves what kind of person we are becoming through our life and work choices. Have we given in to a set of hidden rules that have set us up in opposition against the qualities that are at the very essence and best of who we are? In these moments we pause to reflect and realise that the quality of life for many future generations rests on the leadership decisions that are taken today. This needs to be taken seriously but the passion for change will encourage us to transition to new ecosystems that respect nature and people in the way we create and extract value.

The Carbon Majors Database that records data on historic greenhouse gas emissions have found that 71% of emissions since 1988 and 52% since the start of the Industrial Revolution in 1751 have been traced to just 100 companies. These include organisations such as Exxon Mobil, Shell, BP, Chevron, and BHP Billiton. The future of our existence on planet Earth is in the hands of 100 business leaders with greater power and opportunity than democratically elected governments.

How do we respond to this and how do the leaders of these businesses respond? Our experience is that shouting from the other side of the Zambezi, from the elephants to those still on the bank, 'come here or else', is unproductive. Similarly, the NGOs seeking to jump on genuine but at times clumsy communications of authentic sustainability intent can clip the wings of pioneering CEOs. These are all well intended actions but will have the effect of driving leaders back into Day 1. Compare this with the leadership qualities of Wilberforce who went to where leaders are, sought unity around the need for moral conviction to deliver change and deployed servant leadership in guiding leaders as they embraced the danger and opportunity of the transition zone.

For the avoidance of doubt, we are not for one moment de-prioritising the need for genuine commitment from the leaders of these 100 companies, but we need to make their journey less treacherous and more likely to succeed. This will be difficult if we see them as an enemy that needs to be defeated. Yes, some will fail but we need heroic stories of success. We need their infrastructure, their engineering skills, their capital, and their leadership, to make the pivot into Day 2.

Hitting leaders over the head with a proverbial stick will be less productive than guiding them to the solutions that can solve the world's most pressing challenges. During his tenure as chairman of Canon, the late Ryuzaburu Kaku led the company on the guiding credo of *kyosei*, a spirit of cooperation in which everyone works together for the common good. He foresaw that the world would need the world's largest corporations to work together harmoniously: 'When practiced by a group of corporations, *kyosei* can become a powerful force for social, political, and economic transformation' (Kaku, 1997, p. 55). Many of the top 100 companies are moving into the transition zone taking investors with them but, as we mentioned in Chapter 2, they face industry and political elite strongholds that require extraordinary courage and wisdom.

Courage in the Canoe

As leadership expert and co-founder of GiANT Worldwide, Steve Cockram puts it:

> We need courageous leaders. These are people who are prepared to do the right thing even when everything within them wants to say, 'I've got three years to go. How do I keep my nose clean so I get a good pay off and go onto the next thing, and leave the mess for the next person?' That's the call. Yes, it's understandable how, after years of working in existing, old-era strongholds, some leaders have become highly political and deeply cynical. However, we need leaders who have the courage to do the right thing and, ultimately, remember why they went to lead in the first place.
> Cockram, 2022

It's this combination of heroic leadership and servant leadership that will be required to steward organisations into Day 2. As we shall see later, these qualities are not always in one individual. Often the tension between risk, compliance, and value creation is held within a leadership team, but we will also explore later the key role of 'The Guide'.

Guillaume Le Cunff highlights how, for him, this is a balance he seeks to continually strike. 'I believe that anything related to sustainability or any commitment to global challenges requires a bridge with the business. The challenge for all involved in business is to connect the two. Let's make sustainability both an impact and a business topic and eventually an opportunity for all.'

For Le Cunff, doing business is a continual process of oscillating between these two goals, reconciling sustainability and impact with making money. In many ways it is as simple as that, and artfully navigating these challenges is what drives Nespresso's sustainability goals. For a business such as Nespresso, embedded in one of the largest multinational corporations in the world, navigating the seismic transformation that will eventually be required to truly realise the Day 2 opportunity of total value creation is a process. But Le Cunff understands that sustainability must be embedded in all decisions coming from the very top. He is pragmatic and, in many ways, unsentimental about the approach of constantly turning over the questions, what do I want to do (relating to impact and sustainability), how can I do it and how do I create value?

Whether you are Le Cunff exploring the possibilities of total value creation, Polman resisting the myopia of short-term reporting, or Mastantuono developing and deploying the tools and analytics for the adventure, you are pioneering and pressing into the excitement and hope of Day 2.

The truth is though, that the world's biggest corporations are often being led by 'play-it-safe' leaders. They have steadily made it to the top, and understandably feel safe in their stronghold, with a future that is set and secure. But 'safe' is no longer safe. We need the pioneers.

In our experience working with pioneering leaders who resolve to move towards Day 2 strategy in their organisations, success comes down to three mindsets. First, to be open-minded and have a deeper awareness and perceptiveness about the wider world in which their organisation operates. Second, to have a clear understanding of the reality of now and appreciation of the risks and the required speed and scale of change – in short be prepared to go for it. Third, to be comfortable and courageous enough to leave behind old ways of doing things – to leave the stronghold or redesign it.

These mindsets are not exceptional, possessed only by a few. In fact, the diligent, determined, competitive passion that makes leaders aspire to lead in the first place is just what is required for leadership today and in the future. But this needs to be balanced and guided with humility, pragmatism, true authenticity, and hope.

The Enterprise of Day 2

Purpose, as we have explored elsewhere, is a 'muddy' term often overused and under-realised, but striving for its true meaning for your business and working towards this 'North Star' is essential for the leader exploring Day 2. For us, Purpose is simple: it's the positive impact you want to create from the actions of your business. Authentic Purpose is aspirational and directional and often driven from within leaders themselves as they embrace the dualities of leading for the new era and connecting the 'dots' of who they are. For some, getting to this place may seem incongruent with the hustle and hard edge of business. But view purpose as 'soft' or optional at your peril. Profit will always be a core driver, but leaders who confidently and unapologetically strive for the balance between profit and purpose will be the ones driving business success in an uncertain future.

Leaders for the new era who will succeed in embracing the Day 2 adventure, will engender a spirit of entrepreneurship. As we read at the top of this chapter, we don't know how to get where we need to go, we don't yet know the route, but the best leaders will be those who embrace this uncertainty, experiment, and 'play'. They will make mistakes, they will be attacked for these mistakes but their convictions will instil a resilience galvanised by them understanding why they matter.

As we learned earlier, fashion leader Geoff van Sonsbeeck was already leading successful maternity fashion brand Isabella Oliver when he and his wife discovered that every second a truckload of garments is going to landfill or incineration somewhere in the world. As this statistic sunk in, they knew that they couldn't return to their business and attempt to do some tinkering here and there; they had seen that the whole system was broken and that they had to attempt to deconstruct the model as they knew it, and to change the whole industry. He told us:

> No small challenge, right? I think it's an opportunity and I love challenge. I love change. If you start unpicking it and literally unlearning what you know, you start seeing a better logic that is doable. And, of course, you engage your team, you start engaging your suppliers and you create excitement around a project that creates its own momentum.
>
> Van Sonsbeeck, 2022

As intimidating as that process seemed, Van Sonsbeeck's entrepreneurial spirit, and ability to sit with uncertainty sparked excitement and energy. Baukjen are leaders in sustainable fashion, collaborating across the sector, developing new technologies and business models, cultivating different types of stakeholder relationships, and growing a business fit for a sustainable future.

Reflecting on this process Van Sonsbeeck shares:

> *It's much more fulfilling. It connects the dots across teams and suppliers in a very natural way. It is interesting to reflect on why we held ourselves back in the past. Why did we think so differently? It is so natural to be very clear on the purpose that unifies the whole supply chain and how people think. As a manager and as a CEO, my management is so much easier because all I'm asking for from my team is to do what's right. My job is mostly about reminding and seeing how I can help them remove some of the blockers. Of course, at no point is there an illusion that we're no longer looking at profits, profit is absolutely vital. But we'll be making profit by putting the spotlight on where the costs and the opportunities are and then seeing how we can navigate this, and we have to make trade-offs. We can't do everything today.*

The leaders that people will want to follow in the future will be those who are bold and courageous, unafraid to embrace the world in all its chaos, to make difficult decisions and seek solutions through innovative, entrepreneurial thinking, seeing the fun in the adventure. But that is only half of the story. These leaders will also need to be able to cultivate calm in the middle of the mess, cutting through the 'noise' to extract the essential from the complex, to confidently challenge strongholds with humble maturity and inspire those who follow them to envision a better world.

Humility and Self-Reflection

Humility isn't always easy to define but is easier to identify in others. Leaders who display humility are those ready to acknowledge they don't know or hold all the answers, they admit their shortcomings, lack of understanding, or control over situations. They don't lead through their insecurities.

Humility allows us to lean on others who possess more knowledge or understanding than us, who may be better positioned for a particular role or task. Sean Simpson is the Founder and Chief Scientific Officer at LanzaTech, based in Chicago and with global operations in China and Europe. In his interview, Sean emphasised the simplicity of it all. 'Mistakes are made continuously. Have sufficient humility to know that you are going to make mistakes. You've got to constantly evaluate things. As soon as you realise it's a mistake, back straight out and don't feel bad about it.' And yet we know the world is filled with leaders unwilling to admit fault or culpability, unable to reflect and reframe their views of the world.

Humility is simple, but hard to achieve, especially if we have spent our lives being encouraged to achieve, to hustle, to fight for position, seek promotion, ignore our 'soft skills', and push for the best. We might consider that ambition doesn't fit with humility, that being authentic and transparent will make us seem weak. On the contrary, the leaders who will help us to navigate through and solve the world's biggest problems will display authentic humility alongside fierce grit and determination. Humility is not weakness; it is often a display of the greatest strength.

We can begin to cultivate humility through self-reflection. The ability to shine a light on ourselves, step out of our worldview, reframe our ideas, perspectives and understanding is crucial for leadership in this moment. Humility is knowing your limitations alongside your super-power and an ongoing learning of how you understand, process, and engage with others. It requires us to step back and view a situation through a different lens, allowing yourself to be unsettled and your perspectives shifted.

Humility contains the ego which, unchecked, can either lead us into foolhardy risk-taking or a false certainty that we're right because of competencies and practices that have given us success in the past. Wise leadership in liminal spaces combines the characteristics of humility with boldness, and balances rational and experimental thinking to develop an exceptional understanding and flexibility.

Dame Polly Courtice, Emeritus Director and Senior Ambassador of CISL, is struck by how much knowledge there is in business leaders and the business community 'but how un-joined up it is':

> People have been encouraged to think so narrowly about the only bits that matter to them. But when we present that big piece on what's going on in the overall system and join up all the system pressures and trends, they're often shocked. I can't believe how many amazingly well-educated people are still shocked at the interconnectedness – and therefore the consequences – of what's going on in the world.
>
> Courtice, 2022

For many leaders, engaging fully in the sustainability challenge requires a reframing of their worldview. Sometimes, even acknowledging that strongholds exist and that we are in one requires us to reflect deeply, show humility and acknowledge the need to consider new ways of doing things. Dame Polly observed many of the top leaders on her courses, those who rarely have a moment to sit back and reflect. She sees how when they learn about the deep system pressures and interconnectedness, instinctively they want to move quickly to the what and the how, to make a plan, and get the job done. 'The really interesting leaders to me', she says, 'are those people who are willing to step back and recognise what's happening, to think about the change they want to lead from a long-term perspective, and work backwards from there.'

Something amazing about this moment is that in the unsettling of systems and the search for longer-term solutions, leaders are being required to bring their full selves to the challenge; to connect the different parts of their lives and appreciate how it can and should fit together. For many, this is a departure from everything they have experienced in their work life, but, for most, it also seems right.

The leaders who will thrive as the world shakes and strongholds crumble will be those who know what drives them deep down, and who allow this understanding to filter into their work, their ambition, their vision.

They also know when to draw on the insight and expertise of others.

ESG as an Acronym for Leaders

In a sector more riddled with acronyms than most others in the business world, perhaps we could be allowed to extract the IP asset of the ESG acronym and repurpose it for our Day 2 agenda.

As we've determined, ESG has emerged as the *primus inter pares* of sustainable business acronyms towering over the now vanquished concept of CSR. It is emerging as the preferred framing of the relationship between impact and performance in the investor community and in business more broadly. Many organisations have already created ESG departments, often to strengthen the walls of the stronghold and to comply with the various exigencies of risk-averse investors.

For those leaders and organisations excited by the opportunities of choosing value creation and the adventure of Day 2, we suggest an evolution of the ESG acronym as a new, refreshed, and repurposed IP asset that sets out the three components of a successful future combining positive impact with Sustainable Performance:

E for *entrepreneurship*

S for *spirit*

G for *grit*.

We see these not exclusively as qualities of individual leaders but the dynamics of successful teams – groups and coalitions of the willing inside organisations and in partnership between organisations. They apply to people, whole organisations, and platforms of collaboration in equal measure.

Entrepreneurship and entrepreneurial thinking and endeavour is a critical starting point. It requires the courage to move beyond current ways of thinking and to be open to the new. It is about being wide-eyed to the reality of now and seeing opportunities in addressing the global challenges we face together. It is about pioneering, exploring, staying externally focused – on customers, suppliers, and other stakeholders – and finding new ways and new solutions to create common agendas. Whether you are a born Day 2 native or fast mover to Day 2, entrepreneurialism describes the agility, ingenuity, appetite for risk, imagination, and opportunity-focus needed to thrive and drive Sustainable Performance.

The second quality is what we call 'spirit'. By this we mean a number of things. It includes the ability to see further and more broadly and beyond yourself, your needs, and the narrower introspective constraints and considerations of the organisation. It is about openness, the ability to integrate heart and mind in decision-making, to exhibit humility and empathy in our relations with others, and assume positive intent in the stakeholders we work with. It is about imbuing our leadership with a spirit of service within the organisation and beyond its walls – to turn our strongholds into open and inclusive places of joint endeavour and protection. It is the ability to imagine and build projects and narratives of hope to inspire the loyalty of others to a common higher cause.

The third element is grit. Nowhere in this book are we saying that the pivot to Sustainable Performance is easy. If it were, we would see it all around and there would be no need for books like this or guides such as us. It's hard. Not only does it require the ability to leave *terra firma et cognita* and sail (or canoe) to *terrae incognitae* but it requires the resilience to stay on track despite the setbacks on the way. This grit and determination is required to stand up to the nay-sayers who will tell you that the Earth is flat and you must surely fall off its end. Grit will flow naturally to the people who have a passion to do in their lives and at work what they feel called to do. Individuals and teams will require great focus on outcomes, agility and flexibility in navigation and tenacity to achieve the final destination.

Based on the work we have done and continue to do with Day 2 leaders and their organisations, we are convinced that this blend of characteristics will:

- rapidly transition businesses to the model of the next phase of the economy;
- connect with the hearts and minds of customers;
- attract the talent for the future;
- appeal to investors looking for returns and not just business as usual and capital protection;
- gain loyalty in supplier relations;
- inspire change in the wider category; and
- make a meaningful contribution at scale to the most material issues for the business and the specific area where it can make the most relevant and material global topics.

Guides

Writing this book as two white, middle-class, Anglo-Saxon men, we are aware of the trap that many leaders fall into, often without even realising: surrounding ourselves with people like us and advising each other using our frame of reference, our dominant worldview as our guide.

Even if we did not want to take this approach, we must acknowledge that we are also part of a stronghold that may seem practically bullet proof. Our cultural strongholds may be defined by gender or race, patterns of thinking governed by our education, our upbringing, or our access to knowledge and education. It is the privilege that elevates us above others, and once we are up there, the cultural wiring that biases us towards privileging others just like us.

As leaders, therefore, we need to be mindful of this and about the people and teams we surround ourselves with. Who can you put in place that will not only challenge, but inspire and motivate you, helping you to reframe your perspective so that different approaches and fresh insights become clear?

The biggest global problems will, of course, not be solved by business alone. Collaboration within and across sectors, civil society institutions, governments, and NGOs is essential. But increasingly business is being seen as the one institution that has the influence and pace required to pivot quickly enough to deliver the transformational change needed in this 'Decisive Decade'. The pressure on leaders today, therefore, is huge and the stakes of inaction are high, but we must also acknowledge that leadership can be a lonely place. Business leaders best trained in quarterly profit objectives, competitive advantage, and the bottom line, find themselves front and centre in the sustainability debate, called upon to spearhead transformation for an unknown future. Lonely indeed.

The key protagonist in our elephant and canoe analogy was the guide. Steady, clear, and knowledgeable, the guide was a voice of clarity and calm during the elephant charge. They encouraged Stuart and Louisa to reflect after the incident, explaining what happened, and suggested a new approach (canoeing under the elephants) for the next day. Despite Stuart's resistance the guide remained calm and gently pushed them into the unknown. Getting into the canoe was an anxiety-inducing prospect – there was, naturally, risk involved – but with the guide's reassurance it also felt a little bit exciting. The experience of course was a good one. The guide elicited trust and the adventure spoke for itself.

We are reminded of the words of former UK Chief Rabbi, Jonathan Sacks: 'When we have too much of the "I" and too little of the "we," we can find ourselves vulnerable, fearful and alone' (Sacks, 2021, p. 1).

Leaders need guides to navigate a path unknown. As we have said, we are all entering the unknown, but in sustainability there are those among us who know the details and have their eyes on the big picture. They can clearly see the interconnected nature of people and places, opportunities, and pitfalls. They have insight that has become sharpened by experience, and they can see defined routes through the mess; understanding what needs to happen here, to create impact over there.

In their book, *The 100X Leader*, Jeremie Kubicek and Steve Cockram beautifully unpack the relationship between the Sherpa guides and the thousands of climbers who have embarked on the adventure of climbing Mount Everest. They highlight the deep knowledge of and reverence for the mountain and the embedded technical understanding that comes from being born into the Sherpa people. The Sherpa people are 'devoted to helping each other and serving their clients to the highest level', they are 'acclimated to the mountain, because they have climbed it so many times' and 'are now healthy enough to climb easily up and down' – and to help others do the same (Kubicek and Cockram, 2019).

The Sherpa guide is not unlike the guide from Stuart's story, and the sustainability guides described above. These are the guides you need to help you lead in this time of transition – someone you trust who is prepared to go on the journey too, preparing and partnering with you to help you to realise and achieve something phenomenal. The best guides won't sugar-coat the challenge, or suggest an easier alternative, they will be honest and firm, gently pushing you to achieve the best. The best in our sustainability adventure is transformation, and what we term 'Sustainable Performance' as we have heard, a way of operating that unlocks value creation advantages by placing the organisation in the service of an environmentally sustainable and socially just future.

As we have said, embarking on this leadership adventure requires something special. Although we feel sure that every leader who has assets to steward and who has reached a place of success would have what it takes, some leaders ultimately won't be suited to this. The leaders who are unwilling to reframe or refocus their perspectives, to consider a different approach, to take bold, possibly uncomfortable choices, may remain in the remnants of the old era and hold on to what feels safe and secure while it remains. We've shared a practical model based on our experience as guides in Anthesis in Appendix 2, 'Building Organisational Will and Operational Capabilities'.

But maybe the story of dualities is one you are drawn to. Perhaps balancing courage and humility, clarity and complexity, calm and chaos resonates more strongly with you as the leader you aspire to be. It will be hard and unexpected and will likely push you to your limits but, perhaps, it is time to consider whether playing it safe is enough or whether in fact you are ready to move from your comfort zone and face the challenge, the risk, and uncertainty to leave a legacy you are proud of.

References

Ben and Jerry's (2022) Our values and mission. www.benjerry.co.uk/values

Bewley, S. (24 October 2022) Interview with the authors.

Cockram, S. (19 July 2022) Interview with the authors.

Cohen, B., Greenfield, J. and Maran, M. (1998). Double Dip: How to Run a Values Led Business and Make Money Too. Simon & Schuster.

Courtice, Dame P. (18 August 2022) Interview with the authors.

Heinzel, S. (8 September 2022) Interview with the authors.

Iggo, C. (6 February, 2023) Interview with authors.

Kaku, R. (July–August 1997) The path of kyosei. Harvard Business Review.

Kubicek, J. and Cockram, S. (2019) The 100x Leader: How to Become Someone Worth Following. Wiley.

Le Cunff, G. (10 August 2022) Interview with the authors.

Mastantuono, G. (14 September 2022) Interview with the authors.

Sacks, J. (2021) Morality. Penguin Books.

Simpson, S. (18 August 2022) Interview with the authors.

Van Sonsbeeck, G. (8 August 2022) Interview with the authors.

Part III

'Day 2': The Reapproach

In the first two parts of this book, we have set out the 'reality of now' and the enormous pressures facing organisations at a time of transition and transformation between the current (or previous) industrial era and a new emerging era.

We have explored the challenges facing leaders and stewards of assets as they think through the personal leadership and organisational strategic response to these pressures.

Will they seek to protect their current strongholds from the Charge, placing trust in ESG compliance or think differently about the opportunities of a sustainable and regenerative future?

In this last part of the book, we will examine how leaders are already navigating the reapproach and finding smart, creative responses to seize the new opportunities afforded by such disruptive change. We will explore how to create, and possibly redefine, value in the transition as we start to reimagine the design of the new era. We will learn what it means to canoe towards and under the elephants.

Part III

Day 2: The Reapproach

7

Natives

How Some Enterprises Are Built for Total Value

Earlier we looked at the potential outcomes of an extrapolation of today's business as usual for our shared future – both for planetary ecosystems and the well-being of humanity. In this chapter we will introduce the idea of Day 2 natives. These are the kinds of enterprises that put impact and purpose and the needs of stakeholders front and centre of their organisational focus and priority. New enterprises are being born in this liminal and transitional time, turning the challenges of the externalities of the previous era into new business opportunities.

We will also see that this public-spirited and purposeful conception of enterprise is not new but has a deep and rich tradition and that real entrepreneurs are often motivated by solving socio-economic and environmental issues. Profitability is an outcome, not a driver.

We will also start to see how entrepreneurs and leaders who think this way will be reframing the parameters for sustainable organisational performance for the years to come. Such business leaders see far beyond a compliance agenda to find new ways to create value and to reset whole industries in line with the demands of a decarbonised and circular world.

Making Choices

We've looked at strongholds and how they can become strangleholds on our thinking, decision-making, and actions, and the implications for the planet and society of the choices we make. But there is another important aspect to consider when contemplating choice and this is about the need for our work and our enterprise to be meaningful for ourselves and for the wider world.

There are many examples of entrepreneurs and leaders who are driven by this need for meaning and impact. They are typically astute, perceptive, and aware of the opportunities afforded by the disruptions we are experiencing. They are creating new organisations that are built for the new emerging era or, at the very least, are able to capitalise from and contribute to the transition. Let's call them 'native canoers'.

Native Canoers

Native canoer enterprises were born on or for Day 2, founded to solve the key challenges facing the world today and turn them into entrepreneurial opportunities. Given that one of the most pressing and material global issues is solving the climate crisis it is unsurprising that we see so many new start-ups commercialising the opportunities in climate remediation and adaptation. In this chapter we shall showcase some native canoeing organisations bringing new ideas to face the real elephant – which is the one in the room: the climate crisis and its impact on planet, people, and enterprise more generally.

We spoke with the founders of two native canoeing enterprises, Sean Simpson of LanzaTech and Allison Dring of Made of Air. Both companies had been set up to commercialise and scale solutions to the climate emergency, each of them via a specific approach to an industrial sector called carbon capture.

There are currently two classifications of carbon capture technology – CCS (carbon capture and storage) and

CCU (carbon capture and utilisation). The former is a kind of reverse logistics solution applied to carbon emissions, putting the carbon removed from the atmosphere back where it came from – underground. CCS always feels like the kind of logical and very rational solution to the problem that might come from engineers or geologists or geo-engineers. It feels like a technically brilliant idea but challenging to attract private sector capital to invest in burying a waste by-product and any such scaling of CCS will almost certainly require levels of state subsidy or support from the carbon markets.

The second option, CCU, is more appealing for impact entrepreneurs like Sean Simpson and Allison Dring who see captured carbon as a new resource – a raw material that can be converted into multiple useful and valuable applications.

Sean Simpson, who we introduced in chapter 6, founded LanzaTech 2014 and in 2022 it announced a pre-IPO on the Chicago stock exchange with a valuation of US$2.2 billion. LanzaTech is already selling its products and services to companies such as Unilever and L'Oreal. Sean, a native New Zealander, is a scientist and insists that he is not a businessperson. He was always fascinated by chemistry and chemicals as a resource, not as pollution, and he just hates waste.

LanzaTech uses bioagents and sophisticated technology deployed onto smokestacks in the steel industry to capture carbon and other emissions and bio-convert them to a range of different substrates that can be further refined and introduced into new material streams. LanzaTech makes aviation fuel from captured carbon, provides inputs for plastic bottle manufacture as well as for the detergent industry and plans to be able to convert the raw material into alternative polymers. Sean says:

> LanzaTech did not start out as a tech company . . . Me and a group of people were looking for an abundant source of alternative inputs into the energy sector as a viable improvement on fossil fuels. We realised that source was waste and then we realised that it was gas and CO_2 itself. We have to stop thinking about waste and emissions and start thinking about resources and inputs.
>
> Simpson, 2022

Leaders like Sean see nature charging at us in the form of climate change caused by displaced carbon, and they also see how nature provides us with the ingredients and clues to the answers. Sean mentioned how others looked at LanzaTech's installations collecting chemicals from emissions, and remarked how industrial, dirty, and ugly they are. He reminds us that 'biological organisms grow up in chaotic, dynamic and aggressive environments. Where life is thought to have started on Earth is in the thermal vents in the deep ocean where there is billowing carbon monoxide, hydrogen, methane, and carbon dioxide. Total chaos and that's where life started.' For Sean, emissions are life!

Made of Air is a company based in Berlin set up by architects, designers and leaders from the built environment sector who also wanted to use carbon captured from the air. Made of Air supplies materials for buildings and claddings in the construction industry and high-end components for the German automotive sector and even upscale kitchen units. Their approach was different and was based on a mission to ensure that the carbon that is captured needs to be fixed into durables for as long as possible. In that sense, Made of Air is like CCS, but the storage is above ground and the fixing is useful as an alternative to other carbon intense building materials. Made of Air's co-founder and CEO Allison Dring told us:

> *We see ourselves as a climate company first with the mission to reverse climate change through materials. If we didn't do it through materials we would do it a different way. We make and process biochar and get it into a state where it can be used for a wide range of material compounds which we then get into the hands of consumer goods companies and construction.*
>
> Dring, 2022

CCU is an industry sector likely to scale rapidly to drive the decarbonisation imperative. But does that mean these businesses setting up around CCU tech are able to succeed when judged against the 'normal' benchmarks of business – attracting investors, finding customers and returning capital to investors at a premium? Well, yes, it seems so. In 2022 Made of Air completed a second stage capital raise. Allison told us that she had seen the distinction between impact venture capital (VC) and standard VC disappearing: 'Every VC is now into climate and sustainability tech investment. I don't even know what a traditional VC who's not looking into that space would look and sound like.'

But these kinds of Day 2 ventures are not exclusive to CCU or even climate and carbon tech more generally. In our need to restore, regenerate and rebuild the natural capital stocks that have been so depleted through expansion of population, agriculture, and industry, there are multiple new opportunities to commercialise.

Another Day 2 native canoer is Arnaud Lacourt, the founder of New York based Ubees, a company that was originally created to offer the ecosystem service of pollination by deploying bees and hives in agricultural regions, supported by digital technology. After colony collapse disorder where thousands of hives died with the loss of significant populations of bees and other pollinators, farmers across the United States were hiring in hives that were brought in on mobile units to be deployed for the pollination of key crops such as the Californian almond. The value of natural ecosystem services was only truly appreciated when farmers had to procure such services from businesses such as Ubees. Ubees places hives with food producers to amplify pollination and increase yields. Between US$235 billion and US$577 billion worth of annual global food production relies on direct contributions by pollinators (FAO, 2016). Ubees has had no problems in securing the growth venture capital it needs to respond to the growing demand for its services.

In different ways the LanzaTech, Made of Air, and Ubees entrepreneurs are building new companies creating financial capital value from natural capital restoration. They are responding to nature's charge by working with nature to create new sources of value. Such companies are founded on an interesting blend of impact-focused purpose, opportunity scoping, curiosity and commercial pragmatism. It is a compelling blend for the new world that is emerging.

Clear and Focused Impact Purpose

In Chapter 4, we shared how the Dutch confectionery business, Tony's Chocolonely, set out to make '100% slave-free the norm in chocolate' following their five sourcing principles.

Cocoa supply chains, especially in West Africa, where most of the world's cocoa is grown, are characterised by low levels of traceability in the supply chain to farm level, making even basic levels of compliance challenging. Tony's has created a business and a powerful purpose-driven brand to address these issues. By taking such a stance they can also be transparent about the issue of bonded and child labour in their own supply chains by highlighting that their management and control systems for detection and remediation are working. They even launched an advertising campaign to dramatise this point by saying that finding child labour showed that their approach was working.

We see Tony's as an example that shows how clarity of purpose lends itself to a Total Value System model, which, in turn, embraces stakeholder capitalism that drives customer and supplier loyalty and delivers superior business performance. Tony's will have had their challenges en route. They are in the transition zone with all the danger of being outside the strongholds of a category that has followed a well-trodden path in the old era but now face the elephant charge, but Tony's are well on the way to achieving Sustainable Performance.

Native Canoeing Is Not New

But this way of setting up companies with a specific social or environmental mission is not a new phenomenon and should not be reduced to an opportunity to create new value from the climate catastrophe. In fact, the motivation to use business as a vehicle to effect positive impact is almost as old as modern capitalism itself and stretches back at least to the Industrial Revolution in the UK.

Throughout the history of capitalism there has been a motivation, perhaps even a primary motivation to provide goods and services that create or offer social utility. In fact, much of the Industrial Revolution was about bringing new inventions to the world, which would have an enormous impact on the quality of life of people, revolutionising productivity, mobility, and the creation of cheap sources of energy.

Throughout the nineteenth and twentieth century there are myriad examples of entrepreneurs and leaders who were determined to maximise the full social utility of this revolution for people, either for the employees, customers, or the wider world. In the same way as the code of *kyosei* imbued the mercantile culture of Japan, the nineteenth century saw an incredible wave of socially engaged enterprise driven by a Christian group called the Quakers.

Quakers were innovative, entrepreneurial, and diligent but also generous in the way they cared for their workers and shared the significant financial rewards of their companies. From 'Boots the Chemist', founded by John Boot in Nottingham in 1849, and now Walgreen Boots Alliance, to Cadbury, the chocolate brand and now part of the Mondelez Group, the Quakers transformed the British business culture of the time.

Other innovators and leaders at the time reacted to the major 'externality' of Victorian England – namely the living conditions of the new urban populations employed in manufacturing in the fast-growing cities of Liverpool, Leeds, Manchester, Newcastle, Cardiff, and Glasgow. One notable example was Robert Owen, the father of what has become the Cooperative movement. Owen was a mill owner and was horrified by the conditions he saw in the communities growing up around the mill towns. Owen formed a powerful vision for cooperatives in which workers, and even customers, could become co-owners in the companies where they worked or from which they bought goods and services. The idea was revolutionary but it stuck.

The Co-operative Group is still a major grocery retailer in the UK and in different industrial sectors there are cooperatives all around the world. We met recently with one of these, the Mondragon Group, set up in 1940s post-civil war Spain by a Catholic priest, José María Artizmendiarrieta, and headquartered in the Basque Country in Spain. Mondragon comprises 95 cooperatives, 257 different companies and employs 80 000 people with an annual revenue of over €11 billion. In 2022, Mondragon was featured in an article in *The New Yorker* magazine which contrasted the foundational spirit and 'genuine commitment' of the organisation with the more superficial and marketing led approach of Corporate Social Responsibility programmes of large privately capitalised organisations.

In some cases, successful enterprises evolve naturally and almost accidentally out of a motivation to do good in the world. Take for example the story of the Kellogg's company, known for its high-quality breakfast cereals distributed all around the world. John Harvey Kellogg was a medical doctor who ran a sanatorium in Battle Creek, Michigan, USA. He was determined to find the most wholesome and nutritious way of feeding his patients. In this work they tried to harness the power of sunlight. John Harvey and his brother, Will Keith, experimented with various ways of preparing corn in their kitchens. After many failed attempts Will Keith woke up one morning to find that after processing some corn it had hardened into many small flakes. The corn flake breakfast cereal was born but was only ever designed for patients.

The success of the company in the coming years came as quite a shock to the brothers. Their motivation was never wealth. In 1934, W.K. Kellogg donated more than US$66 million in Kellogg Company stock and other investments to the W.K. Kellogg Trust (equivalent to over US$1 billion today). As with other endowments, the yearly income from this trust funds what is now known as the Kellogg Foundation.

This kind of instinct to solve problems and create public good runs deep in leaders who have set up companies. It is, perhaps, found more in the entrepreneurial motivation of the founder and history has shown that it is not always easy to maintain that spirit in successive generations of managers. The care of such public-spirited entrepreneurs evolved into the twentieth century with companies like Anita Roddick's Body Shop, now part of Natura, and Ben and Jerry's, now part of Unilever.

Natives of the Future

In this period of transition from the asset bases and strongholds of past eras and through the liminal space and time we currently inhabit, we see organisations that are being created for the transition and we can also look back and be inspired by the non-conformist leaders of the past. But we can and must also fast forward to a world that will be shaped by new technologies and industries that are barely on the radar of the public or even most leaders right now.

To use decarbonisation as a proxy for the more general array of changes and transitions which we need, we already see Day 2 natives growing up in the aftermath of the emissions of the fossil fuel era. Such organisations might be set up to capture and store or use carbon, or to create and trade carbon 'offsets', like the Climate Neutral Group in the Netherlands. Or they might be pioneers in the value of 'insetting', seeking to reduce and remove carbon by designing nature-based solutions especially in agricultural value chains.

But we are at the threshold of a new world where enterprises are being and will be created that fundamentally shift the relationship we have with carbon. We call this 'carbon resetting'. Such enterprises are designed not to incrementally improve the carbon performance of existing sectors but to reinvent them from the ground up by applying AI and other new scientific advances. One example of a company which does this in the beverage sector is Cana, the pioneer of beverage printing. Cana is built on the proposition that most beverages are water with molecular differences that separate an orange juice from a glass of wine. From 2023 the company is set to commercialise home beverage printing that will allow consumers to prepare their beverage at home avoiding the agricultural, bottling, and distribution stages of the value chain. It points to the current 400 trillion litres of water used to grow ingredients for beverages worldwide and the 543 million tonnes of CO_2 caused by creating, packaging, and shipping beverages around the world. Cana is at the heart of a movement to decentralise the beverage industry's manufacturing and bypass the waste generated in the value chain.

In an interview with Samuel Alemayehu, an Ethiopian-American entrepreneur and investor and one of the co-founders of C1 Ventures that incubates and invests in ventures that use Single Carbon (C1) gases as feedstock, one can soon sense how a carbon reset economy has the potential to disrupt multiple carbon intensive sectors.

> *The implications of these new technologies are immense. Perfect Day is already applying precision fermentation to make zero carbon dairy products but from real dairy and not plant-based alternatives. Synthetic biology has the potential to make many of our current agricultural systems redundant. Boston Metal has new technologies for zero carbon steel and innovation in material substrate inputs has the potential to create zero carbon cement too. One by one high carbon industries will be leapfrogged by these innovations which bring the same quality with zero carbon footprint and at a fraction of the cost.*
>
> Alemayehu, 2022

Compliance and Performance

What is most striking about such organisations – whether recent climate start-ups, future game-changers, or more established ventures, whether capitalised as private companies, recipients of venture capital, or cooperatives – is that they each determine and follow their own unique path. When you read the stories of their founders you find characters with vision and determination. In many cases, as they created their organisations, they were not blinded by being in business, rather they were guided in business by fundamental convictions of purpose and impact. Yet despite not being fixated on traditional models of investment, growth, and financial return, over time their organisations flourished and grew into the ones we know today (or will know tomorrow). Their performance has been a consequence of a non-compliance with the normal way of doing things.

But, before we leave the adventure of Day 2 natives, we should add an important disclaimer. It is not that the power of the venturing in Day 2 means no compliance at all. It's just that an organisation should be focused on creating more value through a Day 2 orientation than the aggregated cost of compliance on Day 1. It's not either or. In this transition period organisations need to balance the demands of compliance with the pull of performance. And when we talk about compliance, we need to be clear that ESG compliance is itself a relatively nascent and woolly idea and it comes in many forms and from many sources. Some of it is regulatory and legally mandated, some is voluntary, but typically represents industry standard expectations.

There are other compliance demands from investors and customers to consider too. It is also in constant flux with new laws appearing all the time, governing important issues related to transparency, human rights, deforestation, use of chemicals, circularity, and more. In this volatile, complex, and multi-faceted compliance arena, wise leaders balance the demands of different stakeholders at the same time as holding on to their core purpose and ensuring that the assets of their organisations are stewarded towards performance against that purpose and to create the necessary financial performance to meet the needs of the respective sources of capital invested in the enterprise in question. Day 2 natives often find that by setting out to canoe among the elephants they avoid the Charge altogether. Their journey itself has the potential to achieve compliance as a function and outcome of performance and not as an end in itself.

References

Alemayehu, S. (6 October 2022) Interview with the authors.

Dring, A. (5 September 2022) Interview with the authors.

FAO (Food and Agricultural Organization of the United Nations) (26 February 2016), Pollinators vital to our food supply under threat. Retrieved 13 December 2022, from https://www.fao.org/news/story/en/item/384726/icode/

Simpson, S. (18 August 2022) Interview with the authors.

8

Journeyers
How Established Enterprises Reconfigure for Day 2

After the previous chapter describing the adventures and excitement of the Day 2 natives and entrepreneurs driving impact and performance, some readers will be thinking sure, it's easy if you start out now knowing what the challenges are and benefitting from the technologies that are being created to address them and aligning your various stakeholders with your new purposeful intent at the inception of your enterprise. Who wouldn't like a blank page with the opportunity to design and build a business from scratch with none of the externality encumbrances of legacy businesses with their assets stuck or stranded in the past? In this chapter we aim to set out some ways in which more established companies are finding their way in the transition and equipping themselves for the reapproach.

Leaders of these companies are important. In Chapter 6 we introduced the potential for heroism within this cohort – we describe them here as the Journeyers. It is clear we cannot rely on the currently relatively small scale of most of the Day 2 natives to address the challenges we face. In the context of the size of the total industrial complex and the transition required, this would be hopeful and naïve. We need rapid transformation at scale, and we need to redeploy assets towards total and inclusive value creation urgently. For this we need large incumbent businesses to pivot.

This is in no way to disparage or downgrade the achievements of Day 2 natives – their vision, courage, foresight, entrepreneurialism, and agility are sources of inspiration to many and theirs are often the heroic stories that are told in the sustainable enterprise community. But there is no denying the challenge of evolving a large multinational corporation (MNC) with its extensive asset base into a successful Day 2 enterprise. Such a transition requires the leadership mindsets described in Chapter 6, an open mind, cognisance of the reality of now and the courage to leave things behind and move on.

But it also requires a set of skills, abilities, and practices of leadership and management that will get the job done, including a determination and conviction to make the change happen and an attention to detail and diligence to ensure the solidity of the assumptions and models for Day 2. All of this must be undertaken in large, global, complex organisations with all of the usual politics, and competing agendas that come with such organisations. Diplomacy and the ability to align and unite motivating forces will be critical.

Journeying Towards Total Value Systems

The pathway from Day 1 to Day 2 is essentially the journey of discovery around the imaginative and strategic redeployment of those valuable assets from a current standard type of value chain to what we will describe as a Total Value System. A summary of the differences between a classic value chain and a Total Value System is set out in Appendix 3.

As with any journey, the experience for leaders will likely be a combination of planning and surprise. In Appendix 1 we have set out what we see as the important elements and phases of a journey towards Sustainable Performance, what we at Anthesis call 'the activator journey'. Like a map it helps in the preparation, orientation, and direction setting. But oftentimes in journeys, and for most people, the excitement is not in the map itself but the discoveries along the way.

The activator journey serves primarily as a planning tool based on decades of experience across our team of guides and explorers. It helps leaders to ensure that the journey is well planned, with a clear destination. But more importantly, it opens up new vistas and horizons along the way and is designed to expand our perspectives and not keep us fixated on the map. Like any other important process of organisational strategic planning, the journey towards Sustainable Performance should be undertaken seriously with the commitment of the leader, the allocation of resources, and internal teams needed to ensure organisation-wide buy in and subsequent follow through. For any journey to have been worthwhile, when we arrive at the destination, we should genuinely find ourselves in a different place and with an altered perspective.

As we consider the journey we shall reflect on four important dimensions of the reapproach. First, the nature of the asset base and the degree to which we can repurpose assets and competences, acquire new ones and generally avoid the risk of stranded assets. The second is more to do with building the necessary foresight to pre-empt the Charge and turn its power into a driver of change. The third area is to reflect on time as a limiter or resource in the transition and, lastly, we'll look at the power of partnership as we build a crew for the canoe.

Repurposing Assets and Leveraging Competences

This book is primarily written as a guide for leaders of organisations who have been entrusted with valuable assets to show how they can reapproach sustainability as a driver of performance. By assets we mean both 'hard' and 'soft' assets. 'Hard' or infrastructural assets include capital, manufacturing facilities, warehousing, extractive plant, logistics capabilities and land. 'Soft' assets include the 'software' around those tangibles: leadership, IP and know-how, the brands, the relationships and experience, and the loyalty in supplier and customer relations.

An important pathway into Day 2 is the idea of repurposing an existing stronghold. Large multinational consumer products companies have been instrumental in shaping modern patterns of consumption with big global brands, heavy promotional and advertising investments, and continuous product innovation.

The companies that made detergents were part of the revolution in domestic hygiene and the reduction of the burden of household chores in the middle part of the twentieth century. They were at the epicentre of the marketing revolution that happened at the same time. In fact, their sponsorship of popular programmes on TV gave rise to the term 'soap opera'. Two of these giants that drive so much consumerism and consumption have fascinating histories and foundational stories. Unilever was founded in 1930 as a merger of companies, one of which was created by Lord Leverhulme, who had established a model community in Port Sunlight in the UK. Leverhulme was a successful entrepreneur and a philanthropist and always had positive intentions for products that would improve hygiene and the communities where these products were manufactured. Over the years Unilever has become known for its brilliance in consumer marketing and building strong brands for consumer loyalty.

Paul Polman, CEO of Unilever from 2009 to 2019, is one of the most inspiring protagonists in the re-injection of purpose into brand assets. Under his stewardship of Unilever, the company committed to the Sustainable Living Plan, which at its heart was a programme to repurpose the company's assets to perform sustainably. The consumer brands that are such a key value driver for the company, and with such strong bonds of consumer loyalty, have been strategically and deliberately repurposed. The strategy is to evolve beyond the current transactional dynamics of consumer-brand value transfer, where brands promise efficacy and imagery, and consumers accordingly open their wallets.

By introducing purpose and impact into the promise of its consumer brand portfolio, Unilever has pivoted the stronghold of consumerism and brand, fully in line with the fundamental promises implicit in many of the categories where it is present – homecare, healthcare, and food. The new consumerism at Unilever is purposeful. Unilever reported that between 2016 and 2017 its purpose-led Sustainable Living brand portfolio grew 69% faster than the rest of the business delivering 75% of the company's growth (Unilever, 2019).

This pivoting of 'intangible' brand assets towards impact and purpose to redefine and cement relationships with modern consumers is supported by research conducted by NYU Stern School of Business and IRI showing that one-third of all new launches in US consumer packaged goods in 2021 were designated by some form of sustainability promise or accreditation (Kronthal-Sacco and Whelan, 2021).

Many consumer product companies have strong competences in consumer insight. We have worked with owners of brand assets to expand the notion of consumer insight towards societal insight. Ramin Khabirpour is a former senior executive at Groupe Danone, the French food and beverage multinational, who before stepping down had been responsible for the dairy business in 25 key markets in Europe. A man of strong values and moral conviction, he was already convinced that businesses had to have a positive impact on society and communities.

Ramin had joined Danone because he was attracted to the foundational philosophy of the company, which had its roots in a famous speech given by Antoine Riboud in France in 1972 in response to the social unrest in France at that time. He proclaimed in that speech that Danone would be a *double projet* by which he meant that the Group would have both an economic and a social objective. This philosophy permeated the organisation and Ramin took it as a source of inspiration and a mandate to ensure that in the markets for which he was responsible, Danone was economically successful and also had a positive impact on societies.

The marketing team at Danone Poland undertook extensive stakeholder interviews and consumer focus groups to gain deeper insight into not only the consumption of dairy products but some of the related cultural and social dynamics surrounding such consumption. As part of the process, they discovered that after the economic adjustments from integration into the EU and Western market economies there was high unemployment, growing poverty and around 20% of children in Poland were undernourished.

As a food company with the resources, financial, marketing, and nutritional assets and capabilities, Danone could make a difference. The key strategic insight was to understand the core competences of Danone as a healthy food organisation and not just a manufacturer and distributor of dairy products. It was decided that Danone could act in Poland to bring the *double projet* to life.

The first major programme to result was a cause-related marketing campaign called 'Share Your Meal!', which promised consumers buying Danone products that for every yoghurt they purchased they would be sharing a meal with a child in need through the partnership with a local non-governmental organisation (NGO) providing nutritional meals for children at school. The programme took off and became a national phenomenon. Hundreds of thousands of meals were provided for children in need, the profile of the NGOs work was raised, and Danone built market share. The campaign has been running for nearly 20 years.

The strategic engagement in addressing the needs of the poor and undernourished in Poland then became a bigger strategic innovation opportunity for Danone. The company formed a partnership with a nutritional NGO and a joint venture partner to produce affordable nutritious powdered milk products, which provided 25% of the micro-nutrients children needed for a price of around €0.10, also giving access to new consumer groups at the base of the economic pyramid in Poland. Ultimately this work led to a global partnership between Groupe Danone and the Grameen social business movement of Muhammed Yunus.

Many large multinational organisations have understood that they are not agile enough to innovate, incubate, and nurture the necessary business models and brands within their organisations. Or, framed more positively, they realise that there is much they can learn by acquiring the brand assets, business models, and impact experiences of other challenger brands in the categories where they operate. A perennial pattern in driving the Sustainable Performance of such companies is the acquisition of more progressive companies. Unilever acquired Ben and Jerry's ice cream and in the deal the founders retained a separate governance within Unilever and a continued commitment to donate 7.5% of pre-tax income to charities.

Danone acquired the social yoghurt brand Stoneyfield and a pioneer of plant-based dairy alternatives, White Wave. L'Oreal initially acquired the cosmetics and beauty care business, The Body Shop, but then later sold it on to a business in line with its ethical origins, Natura in Brazil. SC Johnson acquired the alternative homecare brands Method and Ecover, and the list goes on. Even in our consultancy services category large MNC advisory firms are building their capacity and capabilities for sustainability services by acquiring native pioneers. As in many areas of business strategy the acquisition of useful assets, tangible and intangible, can fast track a company's journey to Day 2.

Foresight and Pre-empting the Charge

Another pathway for large organisations to make the pivot towards Day 2 thinking and Sustainable Performance could be described as the development of higher order vision or 'foresight'. It is always possible to have a bigger vision of what customers, consumers, and indeed wider society want and need in any category.

Recently we worked with a major industrial bread and baking company. While much of the company's production is the provision of affordable baked goods, the project highlighted a range of opportunities to leverage social and environmental externalities to drive product innovation to different ends: to promote gut health; to source and include ancient grains; and to expand the potential of frozen supply chains to prevent food waste. Bread and bakery goods are an established and mainstream market, but the bread category presents many opportunities for big ideas to drive premiumisation or cost reduction.

Another way of thinking about this reapproach is when leadership takes time to identify how a major social, environmental, or economic dysfunction or disruption in the external world will affect the future opportunities of a company. This is about externality innovation, by which we mean understanding the externalities (positive and negative) of our value chains and identifying opportunities there to internalise the issues and make them sources of value and strategic advantage.

The often somewhat inert and unimaginative process of materiality assessment can be a brilliant and productive process to start such an innovation process. Rather than asking what matters to my stakeholders and what is important to me as a business, I can ask what problems matter out there to people, what is important for me, what can I do to solve it, and what are the benefits for the world and for my organisation by doing so.

Treating materiality as a source of relevant business intelligence and as a basis for externality innovation can yield new sources of value creation. Pursuing this route allows leaders to take the full energy of the Charge of the elephant and redirect it creatively and intelligently into a driving force of innovation and value capture as well as impact.

The leadership of Nespresso needed to secure the long-term supply of the rarest highest quality coffee beans for its fast-growing business and discerning customers. The team understood that there were twin material challenges in coffee supply chains.

First, the effects of climate change were beginning to be seen. This was most visible in adverse and unexpected weather events causing crop damage but also changing temperatures were causing higher incidence of plant disease such as leaf rust. Over the longer term, experts – including the Royal Botanic Gardens at Kew in London – were forecasting that the high-quality arabica coffees, which have driven much of the consumption growth in the global market in recent years, would no longer be able to be cultivated as temperatures in producing regions rise.

The second issue was socio-economic. Some of the regions that produce the highest quality coffee were characterised by a high number of very small farms. Such smallholder farms are unable to derive a living income from coffee alone and are particularly vulnerable to economic shocks such as currency rate fluctuations and global coffee price volatility. In 2003 Nespresso created its AAA Sustainable Quality Coffee sourcing programme together with The Rainforest Alliance, initially with 300 farmers in Costa Rica. Today 140 000 farmers in 17 countries are part of the programme. Nespresso and Rainforest Alliance created the first coffee sourcing programme that linked sustainability to quality and understood that the 'sustainability' externalities and risks needed to be addressed to ensure resilience and durability in the value chain for the producers and Nespresso.

The programme was never conceived as a compliance project but always as a driver of the long-term performance of the Nespresso business, realising that loyal and direct relationships with the producers of the essential raw material for the business were fundamental in guaranteeing the promise to Nespresso consumers of excellence in every cup. In our interview with him, the CEO Guillaume Le Cunff set out that for any premium or luxury brand it is critical to spend time in the supply chain and build relationships with farmers: 'Because without a supply chain there is no brand proposition anymore.'

The Nespresso AAA programme has evolved into a flexible platform for innovation and piloting new ways to test and scale solutions to the range of externality pressures in the coffee industry. To address the key material topics of climate change and socio-economic exclusion Nespresso has launched some of the most progressive and pioneering initiatives in the sector. Before most other companies it spearheaded agroforestry, realising how tree planting on farms could build climate change resilience and create new sources of farm income. It has also partnered with the Fairtrade organisation to invest additional premiums to provide protection to smallholder farmers, including crop insurance, retirement savings schemes, and the testing of a Living Income Reference Price model.

The most eye-catching of these innovations, which most vividly demonstrates how negative externalities can become sources of value creation, is Reviving Origins. In this project Nespresso literally journeys to some of the toughest regions of the coffee producing world (post-conflict South Sudan; parts of Colombia shortly after the peace settlement; post-hurricane Maria Puerto Rico) and works with NGOs and communities to revive coffee cultivation. The products that are marketed to Nespresso lovers promise rare and unique flavour experiences with an

enormous positive impact on communities that have suf-
fered and been marginalised. These are some of the fastest
selling products in the Nespresso range.

For Nespresso externalities were never just a charging
elephant running at them angrily, but an opportunity to
think and respond and find solutions to the challenges and
turn them into value creation opportunities. Day 1 risks
became Day 2 opportunities.

The issue of supply chain resilience has also been a
driver of the performance of the UK retailer Tesco. The
company has used its corporate influence and positional
power to drive positive impact, at the same time as improving supplier loyalty as a key pillar of its sustained long-
term growth strategy.

Tesco began as a grocery stall in the East End of
London; its first store was launched in 1929. In 2021 Tesco
became the first UK retailer to offer sustainability-linked
supply chain finance to encourage more of their suppliers
to join them on their science-based emissions reduction
programme. In this programme, Tesco suppliers are offered
preferential financing rates via Santander's supply chain
finance platform based on each supplier's carbon data disclosure, emissions reduction targets and progress against
sustainability goals.

This emissions reduction programme follows Tesco taking a leadership position in 2017 by setting science-based
climate targets for its own operations on the more ambitious
1.5 °C trajectory of the Paris Climate Agreement and, more
recently, committing to reach its net-zero climate target in
the UK by 2035, 15 years earlier than originally planned.

Tesco have proactively sought different relationships
with their supply chain in the last 10 years. Looking for
new collaboration models within the value chain ecosystem,

such as with Santander, and using their position to seek business improvements, enhanced supply chain resilience and loyalty, and positive impact. Retailers have a distance to travel, but Tesco is an example of how a large traditional business sector is stepping into the transition zone.

There are also examples of this different way of pre-empting the Charge in thinking about longer-term risk. Digital businesses are so often the architects of the next era. Gina Mastantuono at ServiceNow told us:

> *ESG is recognised as a business imperative and remains a boardroom priority. We have an unwavering, top-down commitment to leave the planet better than we found it. Furthermore, the next generation of talent demand to work for purpose-driven companies. The war for talent is real and attrition is costly. The landscape of upcoming regulations both at home and abroad make it the right time to prioritise ESG management and reporting solutions. Meaningful, purpose-driven work matters more than ever. Companies that resist will be left behind.*
>
> *Research shows that ESG has close ties to enterprise value, risk mitigation, and financial performance over the long-term. C-Suite leaders can't afford not to figure this out. Customers, employees, and investors are holding companies accountable for progress on ESG as much as regulators are.*
>
> Mastantuono, 2022

Given the fact that the tech sector is so heavily dependent on attracting and retaining the right talent, being aligned with the needs, expectations, and demands of the next generation of talent is critical.

One of the biggest risks facing established incumbents in multiple industry sectors is the degree to which their asset base can become 'stranded' in the context of the pace and scale of change. A combination of introspection and myopia – looking out for the interests of the current strong-hold – may lead to an obsolescence of the very assets that create value currently. This is why it is critically important for leaders to develop the organisational capacity to rethink externalities as opportunities and to reconsider the connection between current assets and competences and the solutions to those externalities.

This issue has been referred to earlier when we consider the huge asset base of the current fossil-fuel-based energy sector and the degree to which it can be either repurposed with hydrogen technology or become stranded as other forms of renewable energy circumvent the need for its plant and utilities. The implications of the Day 2 native solutions and the speed at which they may scale and succeed may mean not just stranded assets for companies but stranded industries or regions, with all the ethical issues that entails.

Financial decision-making tends to justify expenditure that favours investments in capital assets in contrast with investments beyond the factory walls that could yield resource efficiency and reuse, and build stakeholder resilience and brand value. Prioritising capital expenditure over operating expenditure runs the risk of seeing highly valuable state-of-the-art plants being stranded in areas blighted by water or raw materials scarcity or even the loss of employees required to ensure the sustained operation of the plant. Only when leadership thinks more creatively about the trade-off between carbon, water efficient, or zero-waste facilities operating sustainably in unsustainable regions and the return of these investments over a longer term and in the context of a 'charge' of externalities will capital be deployed more wisely to truly protect the value of assets.

Timeframes

Another way in which more established enterprises can navigate successfully into Day 2 is to consider the timeframes for the reapproach. Notwithstanding the urgency of many of the issues we are addressing in this book, much of the impact we are looking at will play out over decades, even if investments are made today. Leaders need to be equipped to balance conversations with the sources of capital about the timeframes for the reapproach and the way in which an organisation is prioritising initiatives, building portfolios of pilots and projects and then taking them to scale. In this case the different ways in which firms are capitalised clearly matters.

In Chapter 6, we mentioned Sebastian Heinzel, the CEO of Heinzel, and his decision to join the family business. A combination of a thoughtful and reflective character and a pragmatic manager, Sebastian talked to us about what drove him to put Sustainable Performance at the heart of the Heinzel Group strategy. Partly this was about the responsibility for an established family enterprise. He pointed out that 'a family business brings very different time horizons. It gives you the luxury of long timelines. It's nothing special to think about the year 2050. My father is now president of the board, and I will be in that place by 2050' (Heinzel, 2022).

This allowed Sebastian to take time before he very consciously chose to join the family business and leave his previous career pursuits. It also meant thinking about the long-term development of the business and how it needs to change to be relevant in the future. We supported Sebastian and his leadership teams across the Group to create a new corporate purpose and a new articulation of what the Group seeks to do in the world. We discovered that in the DNA of the company there is a continuous drive to inject new value into old assets. Heinzel has acquired and invested in mills that needed updating with new technologies and created new assets such as Starkraft, producing high-quality strong paper utilising very low volumes of fibre to achieve that strength. It invests in recycling plants in its group of companies, including one of few plants in Europe with the technology to recycle Tetra Pak cartons.

We assembled, together with the Heinzel team, these data points into a new statement of purpose for the company: 'Recreating value for our partners, our people and our planet'. This notion of the recreation of value implies the ability to extract new value from existing assets, value from products that were at end of life in linear chains and upcycle them through reprocessing. This new longer-term purpose combined with the precise articulation of Heinzel's mission – 'We circulate fibres sustainably. We supply our partners worldwide with the sustainable products they need' – presents a framework for further investment into the next chapter of the company's story.

Many companies journeying into Day 2 are having to become adept at the calibration of piloting and scaling. While that in some senses contradicts the important point about the urgency of the challenges and the need to scale, as with all innovation there is a need for testing and piloting and incubating before major investment.

We have worked with many organisations that have set up specific internal labs or even standalone funding mechanisms to encourage and invest in pilots designed to address challenging and complex externality-related issues. It is prudent to work this way, but we are equally witnessing that there is increased impatience both from leaders within the organisations and external stakeholders to see how quickly successful pilots can be scaled at pace. The ability to rapidly incubate new solutions, usually co-designed with stakeholders, learn from them, and rapidly deploy them at scale will be a determinant of the success of companies in the transition period between Days 1 and 2.

Recently we have been working with larger companies grappling with SBTi net-zero ambitions for 2050 to help identify new technologies and ventures with whom they can partner to accelerate their decarbonisation programmes, realising that they need to start now to achieve the required reduction and removal milestones in 2030 and towards 2050. This is particularly important for organisations with large and complex scope 3 emissions universes. In this work we are already seeing interesting hybrid connections between the longer-term horizons of the corporations and the speed and agility of Day 2 natives. We'll see much more of this in the coming years.

Building a Crew for the Canoe

We see in many of the cases set out above that organisations tend not to get into the canoe alone and not even just with the guide. Given the need to gain foresight for 2030 and beyond, to understand materiality and the potential of externalities to disrupt or destroy or become new sources of value, companies need to build partnerships with stakeholders. Sustainable Performance is about co-creation of value and recontextualising organisations into the intricate webs and networks that include planetary eco-systems and societies. Co-creation is essential in the building of Total Value Systems.

Organisations that seek to drive their Sustainable Performance and navigate to a successful Day 2 experience must be able to build teams that become part of the journey. Too often in the past and in the CSR paradigm, companies defaulted to building pre-competitive platforms with other industry players that then morphed into entities that were working together to strengthen industry strongholds.

What is now needed, and is emerging, are much more pragmatic teams and partnerships developed around a shared commitment to address system-level dysfunctions in a given value chain or build resilience and contribute to SDGs in a certain sector. Such partnerships and coalitions are more focused, clearer on the desired outcomes, and with a bias for action and impact.

Nespresso has a Sustainability Advisory Board comprising the leadership of its key partner NGOs, its brand ambassador, academics, and leaders from other relevant institutions. This Board operates more as a team driven by a shared desire for positive impact in the Nespresso value chain and the Sustainable Performance of the company.

All of the work that organisations will undertake regarding asset utilisation, competence building, foresight, materiality, and externality-driven innovation will be facilitated by the building of platforms, syndicates, alliances, and coalitions motivated by and dedicated to collective action for impact. We are still at the relative beginning of this but are convinced that business leaders will become not only navigators and explorers but also diplomats and experts in identifying the right combination of sources and resources to address and capitalise on opportunities. In fact, we even believe that the best of these emerging platforms will become the entities that move into Day 2 and almost a new asset class of collective enterprise. We look forward to serving them as guides.

References

Heinzel, S. (8 September 2022) Interview with the authors.

Kronthal-Sacco, R. and Whelan, T. (2021) Sustainable Market Share Index 2021 Report. New York University Center for Sustainable Business. Retrieved (8 December 2022), from https://www.stern .nyu.edu/sites/default/files/assets/documents/FINAL%202021% 20CSB%20Practice%20Forum%20website_0.pdf

Le Cunff, G. (10 August 2022) Interview with the authors.

Mastantuono, G. (14 September 2022) Interview with the authors.

Unilever. (11 June 2019) Unilever's purpose-led brands outperform. Retrieved (8 December 2022), from https://www.unilever.com/ news/press-and-media/press-releases/2019/unilevers-purpose-led-brands-outperform/

9

Value

A Reapproach to Defining and Creating Total Value

Oscar Wilde, the nineteenth century playwright, is reported to have said that a fool sees the cost of everything and the value of nothing and one of the characters in his plays says the same but replaces the fool with the cynic. The purpose of this chapter is not to define the indefinable – namely, the true and full nature of value for companies and how and to what we ascribe it – but rather to help the fools and the cynics in the debate around sustainability to understand how Sustainable Performance creates value and how we are collectively empowered to decide what value might be anyway.

For any leader to be able to manage the transition from treating sustainability as an ESG cost of compliance exercise towards understanding shifting externalities as both an existential threat and new opportunity, requires clarity on how value could be framed. And while there may be generalities associated with the definition of value created by Sustainable Performance, we can also look at the specificities of certain organisations in order to open our minds to what value might look like or even mean.

Value Chains and Beyond

The way most businesses have come to understand value creation has been determined by a combination of orthodoxy amplified by the conditioning of business school strongholds. In our reference to Michael Porter's Five Forces model on value chains in Chapter 5, we observed the divergence from it and the forces that are promoting the new business models for the new era.

Porter describes succinctly and accurately the way in which any enterprise creates value through the procurement of raw material or services, the processing and refinement of such inputs through primary (actual delivery of the end product) and secondary (ensuring the smooth governance and functioning of the enterprise) level activities. The Five Forces model is logical and helps leaders to understand how and where they can intervene to drive performance and maximise value to the investors who are backing the enterprise. Porter's Five Forces model has been the foundation for much of what business leaders understand about value creation.

In reflecting on the need for organisations to move from Day 1 to Day 2, it has been interesting to return to the Porter model and understand which aspects about it could evolve to be relevant for the kind of value models that work for Sustainable Performance.

Porter's (1985) original value chain system set out in his article 'Technology and Competitive Advantage' defined the primary activities as the value chain of inbound and outbound logistics, operations, marketing/sales, and services. But the value chain is only partially revealed in the model. We know little about the people and the world beyond the portals of the model. In fact, it looks a bit like a stronghold, concerned with its own activities and processes removed from the reality of everyday life outside. It appears introspective, as if the wider world exists purely to provide services to this partial value chain model, namely supply and demand, and to do so at low costs and high price, respectively.

If such an interpretation of the model is valid then we believe that this form of value chain-driven introspection in the context of the reality of now and the challenges we are facing in the future, gives rise to strategic myopia. Our conviction about where the world is heading is that business enterprises exist and increasingly create opportunity in the extent to which they serve the world and not vice versa. By reaching out into the world, being curious about how it works, and finding challenges to address and opportunities to seize, leaders will drive the Sustainable Performance of their respective organisations. An over-obsession with the primary and secondary processes of value creation within the enterprise risks limiting leaders' horizons and diminishes the potential for Sustainable Performance.

As such, we think that the very notion and concept of the value chain can be refined to be truer to the way in which we not only describe but also create value. Porter himself has famously explored this territory, positing the idea of 'creating shared value' (CSV) in his article with Mark Kramer in the *Harvard Business Review* (Porter and Kramer, 2011). At the time, CSV was seen as a more business relevant and performance-driven concept than CSR. Porter and Kramer explain why the concepts of CSR and sustainability do not work for businesses or at least can only go so far in creating value. CSV challenges companies to identify the points of overlap between their value chain and the needs of wider society. In this sense we agree. CSV focuses the attention of leaders on the truly material issues that face them. However, this extension of the value chain to address dysfunction in areas like supply chain weakness only goes so far and starts to look like a repair job for the value chain rather than a deeper exploration of opportunity in the wider world. It's as if the residents of the stronghold have left the castle to go out to make sure that the local farmers who supply them with food are doing sufficiently well to remain farming. The farmers still serve the castle.

If we return to our earlier definition of strongholds, we see the risk of the current value chain definition and approach leading to value chains becoming defensive strongholds that serve to protect the people within and also becoming strangleholds on our thinking and our ability to create new models. Our vision of Total Value Systems lays the foundation for businesses to become a different kind of stronghold, open and inclusive and a place of well-being and prosperity for all stakeholders.

A more accurate way of describing the value creation processes of businesses in the future is to think of them as value creating systems that create total value. A value system describes the way in which multiple forms of value in and around the value chain are unlocked for our investors and for our stakeholders. An important component of this will also be how enterprises finance their working and growth capital and the nature of the value creation contract they strike with investors.

How Day 2 Natives Define and Create Value

Day 2 visionaries create value as an outcome of the purposeful and impact-driven way that they found and guide their enterprises and they will understand the inputs and outputs to value creation differently.

Sean Simpson, founder of LanzaTech, believes it's all about laying fresh eyes on a problem and considering it from a different angle. Reflecting on LanzaTech's ambitious vision, he shares:

> We are trying to build an industrial ecosystem. Environmentalists will say the steel industry shouldn't exist, or if it did, it should be based on renewable feedstocks. That's well and good but there are trillions of dollars sunk into capital in that industry and they will not get rid of that in the timeframe of the climate crisis. So, we need

*to take those emissions, take all the waste carbon from that
industry and turn it into a resource. And so our business
model is to partner with other groups to complete a circular economy where above ground carbon just goes round
in a cycle.*

Simpson, 2022

Sean and the LanzaTech team see value in the waste
emissions of legacy industrial assets and convert this into a
valuable resource for other sectors.

For Allison Dring, co-founder and CEO of Made of
Air, the approach to financing and value creation was bold,
clear, and unequivocal: 'We wrote the planet into our
shareholder agreement. When considering key decisions,
the company will treat economics and the planet equally.
As we brought in new investors this was a key clause.
Everybody needed to sign up to that.'

While Allison admits that their business model supported this bold approach, as 'the more material we sell,
the more CO_2 we're pulling out of the air', she noted that
the attitude of investors towards future focused companies
like Made of Air has dramatically shifted. 'This was not
the old idea of impact investing where Venture Capital discounts from 10× to e.g. 7×, as there was always a little bit
of philanthropy with impact investment in the past. In our
last seed round there was no hint of philanthropy. This is
all about green opportunity' (Dring, 2022). In this case,
Made of Air made it clear to potential investors that by
investing into their enterprise they would be co-shareholders
with the planet and that environmental value was as
important as financial return to investors.

In the same way that we see unusual and unexpected
ways of investing in and creating value from visionaries
looking for impact through enterprise it comes as no surprise

that such leaders find unorthodox and unexpected ways of financing their respective ventures too.

In Chapter 6 we featured Ben and Jerry's story and how time and again it becomes apparent how they think differently. In *Double Dip: How to Run a Values Led Business and Make Money Too* (Cohen et al., 1998), the eponymous founders, Ben Cohen and Jerry Greenfield, dedicate a chapter to values-driven financing. In a witty dialogue between the two of them, as they realise they need more finance to meet the demand of growing numbers of customers, they acknowledge that they are 'not Wall Street people'.

Looking for alternatives to raising the capital they need for a first raise they turn exclusively to the citizens of their home state, Vermont. They warn investors that if they cannot afford to lose the investment, they should not invest the money in their business. Some years later the company launched a second round of fundraising and this time they decided that the other group of stakeholders who might be interested in investing in them was their customers, so they published the share offer 'on pack'. In the case of Ben and Jerry's, consumers who create value become investors who earn value; stakeholders become shareholders and shareholders are stakeholders, blurring perhaps the artificial distinctions between groups that now have a shared interest in the Sustainable Performance of a company.

How Established Enterprises Reapproach Value for Sustainable Performance

In the early days of what was once called the corporate social responsibility (CSR) movement we were often invited to speak on the topic of the business case for responsibility and sustainability, speaking predominantly to large MNCs on how investment would create value. Often when asked if there was a business case for CSR, we would say 'no', as in our experience there is never a proven business case for anything when expressed at that high level of generality. It was like asking: Is there a business case for advertising? Lord Leverhulme, founder of Unilever and cited above, famously said he knew only half of his advertising expenditure was working but didn't know which half.

There were, and are, plenty of cases in which CSR investment delivers a return and where there is a very clear and self-evident case for increased investment, and there are equally plenty of examples where it hasn't worked and where it has even backfired and destroyed value. A 2018 study by management consultants Bain found that 47% of sustainability programmes are considered failures by companies and only 4% achieved their sustainability goals (Nastu, 2018). One can only imagine the total cost of the aggregated investments into these programmes and how much of this money failed to deliver any kind of financial return. It highlights again the need for robustness in designing strategic programmes and the need for integrated Sustainable Performance.

As mentioned earlier, the nature of value creation for companies driving Sustainable Performance comes from unlocking value in both tangible and intangible asset utilisation. Companies that maximise their ROI from investment in sustainability find smart and creative ways to do both. They create programmes that operationalise sustainability in their manufacturing and supply chain base but they also leverage that for further value creation in corporate or brand narratives. Given that so much of the valuation of companies resides in the value of intangibles, this kind of effective leverage of impact through brands is immensely valuable. However, the potential rewards must be set in the context of the potential risks. True value creation here is in the calibration of measurable impact and the leverage of this through the brand.

Recent years have seen an explosion of climate litigation cases. According to the United Nations Environment Programme's Global Climate Litigation report, the cumulative number of climate change-related cases globally has more than doubled since 2015, bringing the total number of cases to over 2000. Around one quarter of these were filed between 2020 and 2022 (LSE, 2022).

The public relations and advertising agencies who have so often failed to understand that calibration between message and impact, are directly implicated, with 369 cases brought against communications agencies since May 2021. This growing wave of climate cases is predicted to increase as activists embrace this tactic as a strategic move to raise ambition on climate issues and punish corporations for greenwashing. So, for organisations journeying into Day 2 it is important to be mindful of this balancing act and the risks and rewards.

All companies and all organisations are different and distinctive. While it is true that operating to the rules and conventions of a particular category can bring about a degree of what Hamel and Prahalad in *Competing for the Future* (1996) describe as 'strategic convergence', nonetheless companies have different DNA, unique foundational narratives, histories, people, legacies, and destinies. Their leaders may also have different levels of appetite for risk and reward and, of course, they are stewards of unique and distinctive asset bases.

Understanding this uniqueness and distinctiveness of each organisation as well as its relationship to the wider world will present insight and opportunities for that organisation's unique pathway to Day 2 value creation. As we saw in the previous chapter some companies will create value through repurposing their brands, others through shifting externalities to opportunities for innovation and others to attract talent. In all cases leaders move from thinking about cost to value, either in the short term or for the longer-term health of the business.

Guillaume Le Cunff, CEO of Nespresso, framed it this way:

> *At Nespresso we want to make each cost of compliance an investment in potential value creation. And we want to make each investment in sustainability a new business opportunity.*
> Le Cunff, 2022

New Ways to Think About and Bring About Value Creation

How will value be defined and created in these new Total Value Systems? First and foremost, leaders will see themselves as stewards of assets that have been entrusted to them for maximum value creation, not just today but for the longer term. In the rapid changes to a more sustainable world if assets are not evolved and repurposed to be relevant and valuable in the future then they risk becoming stranded at a rate unprecedented in history. The alternative is that these assets are maximised and leveraged for total impact and value creation.

As we saw in the previous chapter and the cases of Nespresso and Tesco, sources of new value will be found in a more strategic and business-driven application of the materiality exercise. Often confined to a form of tick box process in the corporate sustainability reporting 'industry' where it seems to have become mandatory to place the materiality matrix at the front of the report, materiality is so much more. Materiality is a hunting ground for strategic exploration and business innovation. Materiality matters not just to the sustainability team but to the CEO, the Board and the C-Suite.

When done well, and when projective in its scope by looking to emerging material trends and developments, materiality should give us much of the insight we need to rethink asset deployment and reallocate resources as we invest in Sustainable Performance. In our work recently for a MNC food company, we worked with the senior leadership team to develop their first Sustainable Performance strategy and we worked together to translate the material topics on the materiality grid into clusters of opportunity areas for innovation and growth.

Another way of considering value is in the nature of the time frames in which the value is created and also shared. Again, the way the enterprise is capitalised is important here. Sebastian Heinzel points to the inherent advantages of family enterprises: 'A family business potentially offers a source of strategic advantage in the way you start to plan resources and think through sustainability as a long-term agenda. Other companies capitalised in different ways are forced into short-termism' (Heinzel, 2022). As we pointed out in Chapter 2, Paul Polman in his early days as CEO of Unilever, asked short-term investors not to invest in Unilever.

There has been much debate about the nature of growth and in certain countries, notably France, the emergence of the de-growth or *décroissance* agenda. We may start to question the notion of growth as the main driver of value. Indeed, society will have to debate the nature of growth when viewing sustainability through the lens of finite planetary resources. But we have already seen examples of companies where more growth creates more positive impact – where growth and impact are coupled. Sebastian Heinzel reflected on the nature of growth as he took the helm at Heinzel Group, seeing two kinds of growth: 'When I was talking with someone who said there is a word for something which just grows and grows for its own sake – and that is cancer. For me the big shift is from growth without a purpose to growth with a purpose' (Heinzel, 2022).

In Total Value System thinking, value is not the same as short-term financial performance. Value has implicit within it not only the KPIs of the current financial reporting cycle but the asset value that has the potential to continue to drive that financial performance over the longer term and, importantly, in the context of pressing externalities and planetary and societal risks and needs. In practical terms if our assets look likely to become redundant, ineffective, or even stranded in the long term then our value depreciates and our ability to drive long-term financial performance dissipates. We see how the valuation of companies is increasingly being screened by analysts for the exposure to those changing expectations and demands of investors related to major material issues like carbon, deforestation, and exposure to human rights issues. In Chapter 3 we highlight that this process will ultimately result in the adoption of impact-weighted accounting.

All organisations need to work to ensure adherence to the multiple impact areas that are monitored by industry analysts. Ten of the world's largest companies experienced this when S&P delisted them from the S&P 500 ESG index for low performance on ESG criteria. In Tesla's case, S&P cited a decline in criteria level scores related to Tesla's (lack of) low carbon strategy, claims of racial discrimination and poor working conditions at Tesla factories and failures in its codes of business conduct (S&P Indexology Blog, 2022).

Another perspective on Total Value Systems is that full value creation is not always captured directly in the accounting methodologies currently used to measure companies. But in their Sustainable Performance work enterprises create value for other groups that goes beyond enterprise value.

Staffed by many ex-McKinsey executives, the not-for-profit TechnoServe has created an organisation that harnesses 'the power of the private sector' to help people 'lift themselves out of poverty'. CEO Will Warshauer describes the ripple effect of value creation in some of the most socially excluded parts of the world – remote rural communities in developing countries:

> *Corporate investments in supply chains in rural areas create value in so many ways. Of course, they strengthen supply chains and build resilience and reduce risk for the private sector investor. But they also catalyse a microeconomic development flywheel where invested cash can circulate tenfold before it leaves the local economy creating value beyond the actual transactional value for the MNC.*
> Warshauer, 2022

Most of the comments above presume that when we talk about value we are talking about financial and economic value. But recent years have seen the rise in significance of value across multiple capitals. Given the emergency we are facing in biodiversity loss, natural capital, its relationship with economics is at the forefront of this fresh way of looking at value. Tom Williams, Senior Director, Nature Action at the World Business Council for Sustainable Development (a global community of the world's leading sustainable businesses working collectively to accelerate the system transformations needed for a net-zero, nature positive, and more equitable future), highlights the fact that 50% of global GDP is moderately to highly dependent on nature, and in reality, the entire economy is somewhat dependent on nature. Without the services provided by nature we would lose approximately US$33 trillion of value and yet we still find it so hard to price the value of nature into our enterprises and economy (Williams, 2022).

Blended Value from Blended Capital

Although most forms of invested capital might conform to a certain pattern of expected financial return, it is important to recognise that there are different forms and sources of capital. Beyond classic forms of corporate debt and equity finance there is also impact capital. There are a myriad of family offices stewarding inherited wealth and often looking for impact as well as financial return. There are huge foundations often endowed by successful entrepreneurs with a specific mission in mind. There are reserves of public capital and international development finance. Each of these forms of capital seek different objects of investment and different rates of return. Philanthropic capital invests in impact at minus 100% rate of financial return. Some development finance institutions, which steward public money, can reduce or defer returns. If money is an essential ingredient in unlocking value, then it is clear money may have many motivations.

In the same way that leaders will build impact teams to join them on the journey to Day 2, they will also increasingly become skilled in the area of 'blended finance'. According to Convergence (2022), a specialist in blended capital structures in developing countries, 'Blended finance is the use of catalytic capital from public or philanthropic sources to increase private sector investment in sustainable development'. The idea is very simple: if different forms and profiles of capital seek varying degrees of return but share common or aligned purpose in their investment then the pooling of that capital makes sense. In one special purpose vehicle or structure we can pool different sources of finance behind a common impact objective. In practical terms this subsidises the risk exposure of the private sector investing in impact programmes and de-risks development funds as they align with the higher levels of transparency and accountability of multinational listed companies. It is estimated that already today around US$166 billion is invested in blended capital structures.

One such structure is Water Equity, which is a fund or series of funds that emerged from the work of the NGO Water.org, founded by Gary White and Matt Damon to address the issue of access to potable water for the 771 million people in the world who still lack such access. The founders of Water.org and Water Equity calculated that with the current volumes of philanthropic capital flowing into addressing the universal access to water ambition it would take many decades to achieve the objective.

Only by mobilising and aligning private sector finance and commercial capital into the mission are we able to accelerate and scale impact. Water Equity operates by offering interesting rates of return to organisations that want to invest in access, such as large multinational beverage companies like Starbucks and Coca Cola. Their funds are matched by capital from Bank of America and United States Development Finance Corporation that discount their expected return to attract corporates. This pool of capital is then invested in micro finance institutions in the developing world that can disaggregate the funding into small loans to households or villages to install water access points. Such MFIs are known for their high rates of return and low defaults. Already since inception WaterEquity has created access to water for 3.7 million people (WaterEquity, 2022).

Sometimes other sources of capital can be critical in investing alongside the commercial private sector especially in supply chains in excluded, remote, and developing regions. To help mobilise private investment in pioneering projects and higher-risk environments, the International Finance Corporation (IFC), the private sector investment arm of the World Bank Group, uses blended concessional finance. Mariana Petrei, Senior Investment Officer at IFC describes one aspect of their blended finance work is to invest in addressing blockages that prevent value creation that then catalyses the flows of follow-on capital, private or public. 'We expect to see significant growth in these blended structures to create the powerful blend of development impact and financial risk mitigation that will be needed in the transition' (Petrei, 2022).

How Do We Value the Value We Create?

While avoiding the philosophical and somewhat existential discussion around why we are creating value it is perhaps interesting to share some thoughts on the proceeds of value creation. There are probably as many perspectives on this as there are people taking time to consider it.

During the journey of writing this book we have had the opportunity to revisit the stories of some of the great entrepreneur-founder-philanthropists of previous centuries, and to get to know some of the leaders and entrepreneurs creating new companies to address some of the big challenges we face, especially in the journey to net zero. We have been given a glimpse of some of the enormous innovations in the pipeline that have the potential to render vast estates of assets redundant and stranded. We have spoken with leaders of large organisations wrestling daily with the challenge of integrating sustainability into performance, working within the current parameters and expectations of investors. We have had the chance to reflect on our own company and its model of 60% employee ownership and 40% impact private equity investment, which has allowed us to bring more like-minded ventures into the Anthesis world. With these experiences, one thing has been clear: that value is seen and valued differently by different types of individuals and organisations. Investing for total value is as intricate and complex a world as any other aspect of enterprise. In all of it we have been inspired by the way so many people see beyond the purely monetary and financial understanding of value and impressed by how Day 2 opens up opportunities to think differently about what value means to us.

While we acknowledge that Patagonia is not the kind of company that is of sufficient scale to make a meaningful impact in addressing the full effect of nature's Charge, it has had an exponential impact in the way it has used its brand to tell meaningful stories that inspire hope in all of us.

In 2022 Yvon Chouinard, founder of Patagonia, announced that he was giving away the equity and future profits of the successful company he founded to a foundation. From now on, Earth would be the only shareholder in the business. In an open letter published on the company's website he wrote this:

> One option was to sell and donate all the money. But we couldn't be sure a new owner would maintain our values or keep our team of people around the world employed. Another path was to take the company public. What a disaster that would have been. Even public companies with good intentions are under too much pressure to create short-term gain at the expense of long-term vitality and responsibility. Truth be told, there were no good options available. So, we created our own. Instead of 'going public,' you could say we're 'going purpose.' Instead of extracting value from nature and transforming it into wealth for investors, we'll use the wealth created to protect the source of all wealth. Here's how it works: 100% of the company's voting stock transfers to the Purpose Trust, created to protect the company's values; and 100% of the nonvoting stock had been given to the Holdfast Collective, a non-profit dedicated to fighting the environmental crisis and defending nature. The funding will come from Patagonia: Each year, the money we make after reinvesting in the business will be distributed as a dividend to help fight the crisis.
>
> Chouinard, 2022

Comparing Day 1 Value Chains with Day 2 Value Systems

As we draw to the end of this part of the book we have described different aspects of the reapproach and defined the differences between a classic value chain approach to enterprise and a new and emerging Total Value System approach – one that is fit for the transitional period and beyond into the new era. In Appendix 3, 'Classic Value Chain Compared with the Total Value System', we set out the distinguishing features between these different models of thinking about value and shared some additional reflections about the different dimensions and parameters.

References

Chouinard, Y. (2022) Earth is now our only shareholder. Retrieved 8 December 2022, from https://eu.patagonia.com/gb/en/ownership/

Cohen, B., Greenfield, J. and Maran, M. (1998). Double Dip: How to Run a Values Led Business and Make Money Too. Simon & Schuster.

Convergence. (2022) What is blended finance? www.convergence .finance

Dring, A. (5 September 2022) Interview with the authors.

Hamel, G. and Prahalad, C.K. (1996) Competing for the Future. Harvard Business Press.

Heinzel, S. (8 September 2022) Interview with the authors.

Le Cunff, G. (10 August 2022) Interview with the authors.

LSE (2022). Global trends in climate change litigation: 2022 snapshot. Retrieved 8 December 2022, from https://www.lse.ac.uk/granthaminstitute/publication/global-trends-in-climate-change-litigation-2022/

Nastu, J. (8 August 2018) Sustainability change is hard but companies slog along, study shows. Environment and Energy Leader. Retrieved 8 December 2022, from https://www.environmentalleader.com/2018/08/sustainability-change-is-hard-but-companies-slog-along-study-shows/

Petrei, M. (4 November 2022) Interview with the authors.

Porter, M.E. (1985). Technology and competitive advantage. Journal of Business Strategy, 5(3), 60–78.

Porter, M. and Kramer, M. (January–February 2011) Creating shared value. Harvard Business Review.

S&P Indexology Blog. (17 May 2022) The (re)balancing act of the S&P 500 ESG Index. Retrieved 8 December 2022, from https://www.indexologyblog.com/2022/05/17/the-rebalancing-act-of-the-sp-500-esg-index/

Simpson, S. (18 August 2022) Interview with the authors.

Warshauer, W. (21 November 2022) Interview with the authors.

WaterEquity. (2022) Enabling growth, scaling impact. Retrieved 8 December 2022, from https://waterequity.org/social-impact.

Williams, T. (20 November 2022) Email correspondence with the authors.

Conclusion

La vie est la somme de tous vos choix. Alors, que faites-vous aujourd'hui?
(Life is the sum of all your choices. So, what are you doing today?)

Albert Camus

While Shell's CEO Ben van Beurden was speaking at a TED conference before the UN Climate Change Conference (COP26) in October 2021, he was challenged by a climate activist also on the stage. Van Beurden was called 'one of the most evil people in the world'. Less than a year later, Shell announced that Van Beurden was to step down, having been in position since 1 January 2014.

Van Beurden graduated in chemical engineering from Delft University and joined Shell in 1983. Perhaps he dreamt about getting to the pinnacle of one of the most respected companies in a powerful and meaningful sector focused on creating enormous wealth from the provision of cheap, accessible energy.

As we discovered in Chapter 6, since 1988, 71% of industrial greenhouse gas emissions are linked to 100 active fossil fuel producers (CDP, 2017, p. 8). Shell is one of these.

We doubt Van Beurden could have imagined when he set out, that on reaching his career pinnacle, he would find himself leading one of the companies deemed to have brought the world to a dangerous precipice. We think it unlikely he thought he would be called out as being 'evil', or 'responsible for so much death and suffering' (Frost, 2021). Or having to address his daughter when she returned from school in tears because somebody told her oil and gas companies were destroying the world and only Greenpeace could save the planet. 'So why don't we give [all of our] money to Greenpeace, papa?', she said. As she was too young to engage in debate over carbon taxes and the responsibility of all governments, consumers, and corporate polluters, Van Beurden replied: 'You have to trust me' (Raval, 2019).

Wael Sawan, Van Beurden's successor, said in the Company announcement: 'I'm honoured to take over the leadership of this great company from him. I'm looking forward to channelling the pioneering spirit and passion of our incredible people to rise to the immense challenges, and grasp the opportunities presented by the energy transition. We will be disciplined and value focused, as we work with our customers and partners to deliver the reliable, affordable and cleaner energy the world needs' (NNR, 2022).

As the strongholds of the old era crumble and we look into the danger of the transition zone towards the new era, there is much we don't know. But we do know that we will need energy, clean energy. Some of the top 100 companies in the group that we referred to in Chapter 6 will likely fail. Some we need to succeed. We need their infrastructure, their engineering skills, and their leadership and capital assets for the world to pivot in the years ahead.

We have a choice. We can see them as 'evil', as the perpetrators of environmental and societal collapse, the enemy that needs to be destroyed, or we can work with them to get us out of the problem that they've participated in getting us into.

And they have a choice. Hitting the fossil fuel off ramp to net-zero will be an extraordinarily tough challenge for

Sawan. The world is looking at Sawan in the same way as Van Beurden's daughter came to her father. As corporations provide us with their 'you have to trust us' rhetoric, many will get called out as lacking authenticity. Others will make genuine attempts but fail. Then there will be those who succeed. They will be amongst the greatest of the heroic pioneers on the Day 1 to Day 2 journey.

We have featured many pioneers in this book. Businesses and their leaders such as Patagonia and its founder Yvon Chouinard. We love Patagonia. They are easy to love. But they are not in the 100 companies referred to above. The success or failure of our planet is not dependent upon the success or failure of Patagonia, despite being a great company.

But the role of all these pioneers is nonetheless vital. What Chouinard and Patagonia do is to bring a voice to the moral and planetary imperatives of our day and provide a glimpse into the business and design frameworks for sustainable performance in the next era. The entrepreneurs of Day 2, and those courageous leaders navigating the most treacherous business transition of many generations, are the modern-day Clapham sect of Wilberforce's day that will inject urgency and catalyse change with speed and scale.

Becoming Unstuck

We hope the message of this book is clear. We face a crisis that will demand a response. The crisis is framed by the breakdown of economic, institutional, and societal strongholds that were thought to be impenetrable in the context of the old era. Strongholds that have been pounded by ever more powerful externalities. The clue to the significance of this is in the word 'externalities'. They are outside our current thinking and strongholds. They are outside the constructs that we have built for safety and security. They are the essential ingredients of any system and they have been neglected. The engineering design of the post-war model is flawed. We have assumed that we can deploy infinite models for growth and prosperity in a linear model whilst on a foundation of finite resources. This has been known about in scientific communities, and at the most senior levels of industry and in government for several decades. But the warnings have been ignored or the consequences have been so unimaginable that it has garnered a culture of denial. Worse still, many leaders have drawn their followers into a fantasy realm masquerading as the truth. This is all to buy time. Time required to maximise the collection of wealth in the storehouses of the few.

Yes, there are elements of the free market economy where, working in partnership with government and public institutions, the system has shown its friendlier face. We see how global trade has been effective at cutting poverty and inequality. There are many aspects of the free market that have the power to do good. But in 2022, the poorest half of the global population owns just 2% of the global total, while the richest 10% own 76% of all wealth; and while over the last two decades global inequalities between countries have declined, income inequality has increased within most countries (World Inequality Lab, 2022).

So, this is a crisis that will demand a leadership response, and within that response we all have choices. We can hunker down, tick some boxes, buy time, and shore up our pension pot. Or, we can become leaders for our time: courageous, humble, heroic, and adventurous. Leaders that can see the short-term imperative of survival but the longer-term moral imperative to ensure that future generations can enjoy a healthy planet on which to prosper. Legacy will be screaming at many of you reading this. So it should. Some of you will find it easier and possibly more inspiring to imagine your legacy, over imagining some future destination. A destination arrived at via a journey that you may have to start but your successor may need to finish.

It's time for us all to become unstuck. Unstuck from any of the business, political, societal systems, and models that have trapped us in an increasingly destructive way of living. For many of you, this reality will have struck you in recent years, possibly even while reading this book. But we've known about this for a long time, while the world has chosen denial as a place to seek solace. In 1989 Margaret Thatcher, Prime Minister of the UK, said the following to the United Nations General Assembly.

> *What we are now doing to the world . . . is new in the experience of the Earth. It is mankind and his activities that are changing the environment of our planet in damaging and dangerous ways. The result is that change in future is likely to be more fundamental and more widespread than anything we have known hitherto. Change to the sea around us, change to the atmosphere above, leading in turn to change in the world's climate, which could alter the way we live in the most fundamental way of all . . . The environmental challenge that confronts the whole world demands an equivalent response from the whole world. Every country will be affected and no one can opt out. Those countries who are industrialised must contribute more to help those who are not.*
>
> Vidal, 2013

This speech given by Thatcher was a day before the Berlin Wall came down. A reminder that the previous era was characterised by different political ideologies. The post-World War II transition provided numerous liminal spaces, resulting in multiple design choices becoming manifest. There is a beauty in such diversity, but the heterogeneous nature of these responses and resultant leadership, is not an option in this liminal space. Unlike at previous moments in history we have a planetary issue, a shared liminal space, that needs global unity, binding commitments, and a confluence of global ideologies. And most of all we need consistent and concerted action.

Perhaps the demolition of the Berlin Wall is usefully symbolic for the breaking of strongholds that will be required for the reunification of people and nature. We have lived too long apart. Just as East and West Berliners danced and embraced in 1989, we should expect that in this moment in history, the rediscovery of the assets of this abundant planet, will also be a source of celebration.

It's time to reframe our definition of success, which has become an increasingly evasive word. Reframed into the reality that an existential crisis such as we face is the greatest leadership challenge we could ever imagine. The leaders today are the first generation of leaders to have the unequivocal truth, and in this short window of opportunity, they are also the generation we rely upon to lead us across the transition and into the next era. This is the frame against which we need to measure success.

We know that's daunting. But if we get it right, we will enjoy free energy; land and oceans will be revived and become productive; and food and clean water will be abundant for all. Our economic models will drive progress and prosperity, creating inclusive societies that are healthier and with less crime, and where the air we breathe doesn't poison us but strengthens us.

This book aims to illuminate the potential blueprint, to encourage the stewardship of the good from the last era across the transition zone, and with wisdom, ingenuity, enterprise, and human energy, to develop the constitution for the next era. But our primary aim is to provide that glimpse of a future that delivers the hope and excitement of a heroic and historical adventure.

Writing this book has taken the authors on a journey and given us an opportunity to reflect on careers that, when combined, number almost as many years as a human life. It has given us a chance to reflect on the choices we have made in our own professional lives and the choices we continue to make each day. As we mentioned at the beginning of the journey of this book, we see ourselves as inherently Day 2 people and every day we seek to find Day 2 solutions to the challenges we face in stewarding the assets with which we have been entrusted. In this way, we aim to create stakeholder value and drive the continued sustainable performance of our enterprise and those that allow us to be their guides.

As we refer to the oft quoted dictum of Albert Camus at the head of these conclusions, we are reminded that our lives are the sum of all the choices we make. Not all of us are free to exercise choice. Those of us who are, by dint of circumstance or determination, have an immense responsibility and opportunity to make good choices. As leaders, citizens, and members of the global human community we can choose to bury the talents and assets of which we are stewards in the ground or hoard them in the treasure rooms of strongholds to comply with the minimum expectations of compliance. Or we can invest them into the projects, programmes, and enterprises that will yield abundant value for our organisations and our world. As we have set out, we still have choices but the frame is set. The broad sweep of history and circumstance is propelling us to a new era where enterprise must serve the changing needs and expectations of the world at large.

And anyway, for all this talk of peril, we should remind ourselves that in our story, in the reapproach, the danger was always in our mind. When reapproached with the right crew, the elephants held no danger, just wonder. Day 2 elephants aren't scary. We are just afraid. The fear is in us . . . but so is the hope. In that spirit we conclude with a quote from Nelson Mandela:

> May your choices reflect your hopes, not your fears.

References

Griffith, P. (July 2017) *The Carbon Majors Database: Carbon Majors Report 2017*. Climate Accountability Institute. https://policycommons.net/artifacts/3085142/untitled/3885955/

Frost, R. (25 October 2021) Watch as 'evil' Shell CEO is confronted by protestors on stage at TED event. Euronews.green. Retrieved 8 December 2022, from https://www.euronews.com/green/2021/10/15/watch-protestors-storm-stage-at-ted-event-to-confront-evil-shell-ceo

NNR. (15 September 2022) Wael Sawan appointed Shell CEO, as Ben van Beurden to step down. NewsnReleases. Retrieved 8 December 2022, from https://newsnreleases.com/2022/09/15/wael-sawan-appointed-shell-ceo-as-ben-van-beurden-to-step-down/

Raval, A. (27 September 2019) Royal Dutch Shell searches for a purpose beyond oil. *Financial Times*. Retrieved 8 December 2022, from https://www.ft.com/content/45a9b82e-df73-11e9-9743-db5a370481bc

Vidal, J. (29 April 2013) Margaret Thatcher: An unlikely green hero? *The Guardian*. Retrieved 8 December 2022, from https://www.theguardian.com/environment/blog/2013/apr/09/margaret-thatcher-green-hero

World Inequality Lab. (2022) *World Inequality Report 2022*.

Appendix 1: The Anthesis Activator Journey

Every client challenge is different, and we approach any complex problem with a flexible, common, problem- and solution- neutral design process to guide client teams from their initial question, into a fully working, great quality solution for their organisation every time.

Phase 0 - FRAME   Phase 1 - ANALYTICS   Phase 2 - SOLUTIONS   Phase 3 - IMPLEMENTATION

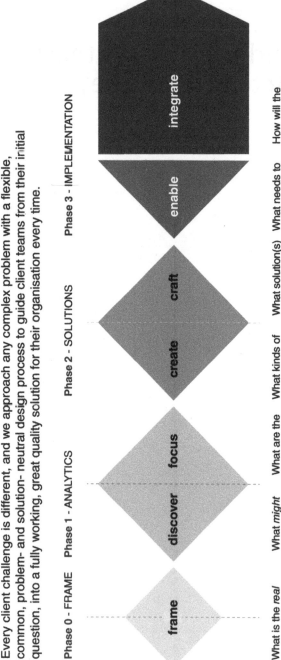

frame	discover	focus	create	craft	enable	integrate

What is the *real* question, problem or outcome, we need to solve for?

What *might* we need to know to solve for this challenge?

What are the implications of what we have learned?

What kinds of solutions *could* work, given what we have learned?

What solution(s) will work best for the client, which they will commit to?

What needs to be in place for the client organisation to implement?

How will the solution become part of the client's model?

Anthesis

Appendix 2: Building Organisational Will and Operational Capabilities

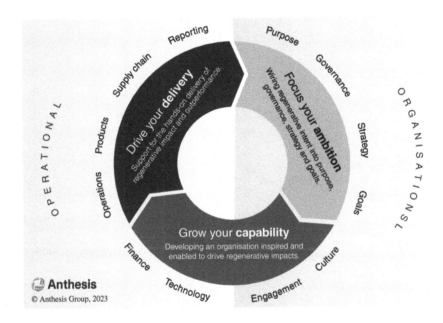

Appendix 3: Classic Value Chain Compared with the Total Value System

BASIS OF THE MODEL

Classic Value Chain	Total Value System	Reflections
Linear	Circular	Value chains are *not* isolated entities.
Extract, manufacture, dispose	Retain primary resource in a continuous production loop	Businesses exist in rich, dynamic systems incorporating society, nature, policy, economy, knowledge, culture and more.
Value understood as essentially investor reward.	Value understood as multi-faceted.	Towards impact weighted accounting
External world	*External world*	
Considered narrowly.	Considered more broadly.	
Inbound and outbound relationships (i.e. suppliers and customers) and related to cost and price.	As per classic value chain but additionally resources, externalities, positive/negative impacts, policy.	
	Connecting and collaborating with stakeholder networks.	
Timeframe	*Timeframe*	
Value creation often limited to near term fiscal reporting cycles.	Value creation is longer term.	

DESIGN OF PRODUCT OR SERVICE

Classic Value Chain	Total Value System	Reflections
Designed to leverage cost of supply vs price to customer through cheapest available sourcing.	Designed to leverage cost of supply vs price to customer by designing for reuse, minimised waste, pricing in external impacts, retaining primary resource in a constant production loop by retrieving product at end of use.	Changing cost profile across business functions that require a different accounting/investor lens.
Little producer responsibility for the cost and impact of waste.	AND	Total Value Systems over the medium/long term will likely be the only viable model when all externalities are built into the costs of company accounts.
	Products and services designed to intentionally and specifically solve social or environmental issues.	

(Continued)

LOGISTICS

Classic Value Chain	Total Value System	Reflections
Forward. Designed for linear, with assumption of high levels of waste at end of life.	Reverse. Designed for circular, with assumption of no end of life, zero waste and 100% reuse.	Shifting large multinationals from forward to reverse logistics will be disruptive.
All business functions (technology, ERP, stock inventories, incentives, talent attraction and retention, facilities, distribution networks) built to leverage buyer power over supplier power to press down on supply side costs and get maximum product to market without considering end of life.	All business functions (technology, ERP, stock inventories, incentives, talent attraction and retention, facilities, distribution networks) realigned to retrieve everything after point of sale.	Requires industry and sectoral level collaboration and new generation of public-private-partners to achieve the volume, scale, and acceleration required.
	Example: Sales people incentivised against the percentage of product sold that is retrieved.	Opportunity for new entrants or reverse models running alongside forward models as businesses switch from one to the other.
Resources vulnerable to commodity fluctuations.	Minimal requirement for primary resource. Resources contained within the value chain and less vulnerable to commodity markets and price volatility.	Transition will be accelerated by intelligent technology solutions for tracking products through the value cycle.

SUPPLY CHAIN

Classic Value Chain	Total Value System	Reflections
Maximise bargaining power of buyers over suppliers.	Engage and invest in suppliers as key stakeholders/partners.	Shifting supply chains from source of reputational damage to source of strategic advantage and loyalty.
Create dependency mechanisms and other tactics to increase leverage potential.	Seek loyalty over dependency.	
View as a commodity that is disconnected from the responsibility of the primary activities.	Recognise the value in the supply chain story as an underpin to the brand story in an age of transparency.	
Focus on cost reduction as the primary source of value creation.	Recognises the emergence of alternative forms of capital and influence in the supply chain such as the creation of carbon assets by suppliers that can be sold across sectors.	
Value travels up the value chain hierarchy.	Co-create value and expect the value to flow through the value cycle for the benefit of all stakeholders.	

(Continued)

SUPPLY CHAIN

Classic Value Chain	Total Value System	Reflections
Assume that influence is contained in the relationship between buyer and seller.	Seeks to align with the values of customers in their buying decisions.	
Lack of/limited awareness of the potential growing constraints on supply for environmental or socio-economic reasons.	Open to innovation opportunities which derive from supply chain transparency.	

COMPETITION

Classic Value Chain	Total Value System	Reflections
Focus on competitive rivalry in a win:lose model.	Focus on finding uncontested space, often through exploring pre-competitive opportunities within and across sectors.	Materiality (and other) tools can help structure a sustainable performance project portfolio within an organisation.
Often leads to a zero-sum game.	Opportunity to position as prime mover in a pre-comp space = 'pre-competitive advantage'.	Through this, ownable 'strategic advantage' and 'pre-competitive' projects can be identified.

BRAND DIFFERENTIATION

Classic Value Chain	Total Value System	Reflections
Driven by:	Driven by:	Ultimately the role of the brand shifts from badging products for customer reassurance to signalling impact to inspire customer engagement.
Commercial proposition	Purpose	
Internal operational asset leverage	Deeper supply chain impacts	
Basic transactional category table stakes	Surprising stances on issues that matter or standout	
Make-believe stories	Truth, authenticity and substance	
Leading to strategic convergence in category and erosion of brand leverage potential and lower prices.	Leading to stronger equity, loyalty and price premiums and value capture.	

(Continued)

END CUSTOMER

Classic Value Chain	Total Value System	Reflections
Assumes the end customer will continue to celebrate cost reduction as primary reason for purchasing choices of one product over another.	Adheres to the belief that customers increasingly expect responsible governance towards the environment and society and the desire to empower and reward this responsibility through the way they spend their capital. BUT Respects cost and achieves cost efficiency from e.g. waste reduction and energy efficiency.	Opportunity to drive eco-efficiency cost savings that may be passed to end customer or redirected to value add impact proposition relevant for end customer. A way of future proofing brands and maintaining future cash flows from brand loyalty. In B2B create relevant assets to align with B2B customer sustainability agenda.

Index